Biochemistry

BIOCHEMISTRY

The Little, Brown Medical Review Series

Paul Jay Friedman, M.D.
Blodgett Memorial Medical Center
Grand Rapids, Michigan

Little, Brown and Company   Boston

To Ralph B. Friedman, my father and the best teacher I've ever had, and Diane A. Friedman, my wife, who provided the encouragement and assistance I needed to complete this book.

Preface

All the atoms of the earth bear witness, O my Lord, to the greatness of Thy power and of Thy sovereignty; and all the signs of the universe attest to the glory of Thy might.

GLEANINGS FROM THE WRITINGS OF BAHA'U'LLAH

This medical review text was written to explain the fundamentals of biochemistry to medical students and to cover thoroughly the topics in a standard freshman medical biochemistry course and in the National Boards (Part I).

When I was a medical student, I found that most existing texts were 800–1200 pages and included so many diverse facts that they were overwhelming. Those texts did not focus enough on the concepts most relevant to medical biochemistry. The emphasis on medicine was limited, and often those texts did not relate biochemical reactions to organs and diseases.

I have designed *Biochemistry* to make learning as it should be—enjoyable. People learn best when they ask questions and then seek the appropriate answers. Each chapter comes equipped with a set of problems along with their step-by-step solutions. By sharpening their problem-solving skills students develop a comprehension of biochemistry that obviates the need for thoughtless memorization on the night before an exam.

No lecture series is needed to supplement this text. In fact, it assumes no previous knowledge of chemistry.

I thank Sandra Wells Ludwig for her services as medical illustrator.

P. J. F.

Contents

Biochemistry

Structure of Amino Acids

1

Structural biochemistry forms the foundation to the understanding of metabolic pathways. To dismiss the study of structure as unessential and dull is as senseless as refusing to learn anatomy. When we separate the chaff—in this case, the superfluous structural details—from the grain—the essential structural features—we are then left with a digestible foodstuff.

All living organisms contain α-amino acids, and all α-amino acids share the same structural backbone:

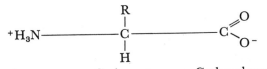

α-Amino group α-Carbon atom α-Carboxyl group

In other words, all α-amino acids possess three common features:

1. They have an α-carboxyl group. The α denotes that this group binds to the central or α-carbon atom.
2. They possess an α-amino group.
3. They contain a side chain, or R group, that is bound to the α-carbon.

Problems 1–6

Which of these structures are α-amino acids? (Turn to the end of the chapter for the answers.)

1. $^+H_3N—CH_2—COO^-$

2. $CH_3—CH_2—COO^-$

3. $HO—\langle\bigcirc\rangle—CH_2—CH—COO^-$
   $\qquad\qquad\qquad\quad |$
   $\qquad\qquad\qquad\ ^+NH_3$

4. [chemical structure: a pterin ring system connected to $—CH_2—N(H)—$ benzene ring $—C(=O)—N(H)—CH(COO^-)—CH_2—CH_2—COO^-$]

5. $^+H_3N—CH_2—CH_2—COO^-$

6. $^+H_3N—CH_2—CH_2—CH_2—COO^-$

You have now proved for yourself that it is quite easy to identify a molecule as an α-amino acid. Your next task is to classify twenty of the common, naturally occurring α-amino acids. *In general, you need not memorize the exact structure of each R group;* knowing the proper classification is sufficient. You should also be familiar with the three-letter abbreviation for each amino acid.

Each amino acid may be classified as *acidic, neutral,* or *basic,* depending on the charge on the R group at pH 7.0 (see Ch. 2). Acidic R groups bear a negative charge at pH 7.0 because they are strong proton donors. The two acidic α-amino acids—i.e., those with acidic R groups—are *aspartic acid* and *glutamic acid*:

Aspartic acid (Asp)

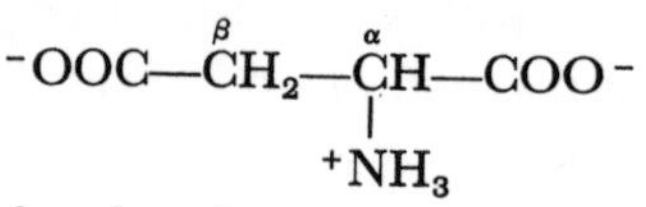

$$^-OOC\overset{\beta}{-CH_2}\overset{\alpha}{-CH}-COO^-$$
$$\underset{^+NH_3}{|}$$

β-carboxyl

Glutamic acid (Glu)

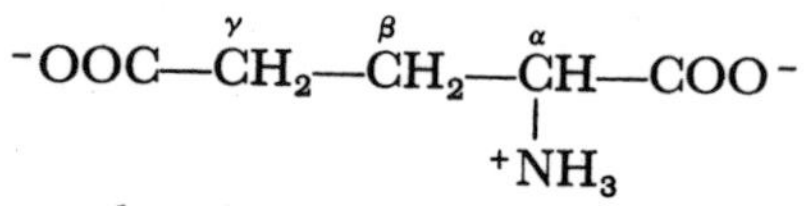

$$^-OOC\overset{\gamma}{-CH_2}\overset{\beta}{-CH_2}\overset{\alpha}{-CH}-COO^-$$
$$\underset{^+NH_3}{|}$$

γ-carboxyl

Glutamic acid differs from aspartic acid only in the number of CH_2 groups contained in its side chain. Each acid carries a charge of minus one at neutral pH.

Basic R groups carry a positive charge at pH 7.0 because they avidly bind protons. The three common basic α-amino acids are *lysine, arginine,* and *histidine*.

Lysine (Lys)

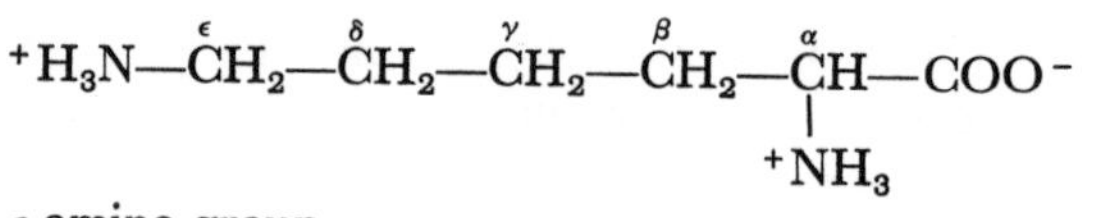

$$^+H_3N\overset{\epsilon}{-CH_2}\overset{\delta}{-CH_2}\overset{\gamma}{-CH_2}\overset{\beta}{-CH_2}\overset{\alpha}{-CH}-COO^-$$
$$\underset{^+NH_3}{|}$$

ε-amino group

Arginine (Arg)

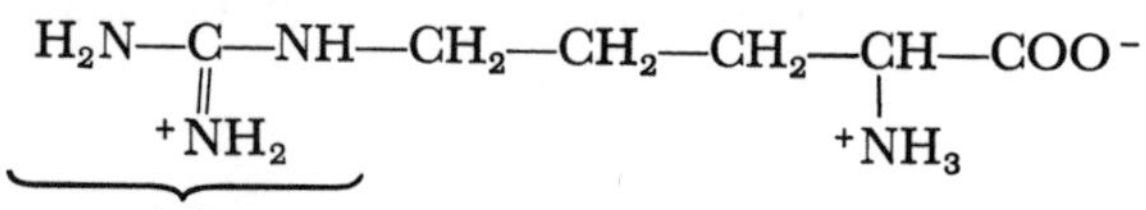

$$H_2N-\underset{\underset{^+NH_2}{\|}}{C}-NH-CH_2-CH_2-CH_2-\underset{\underset{^+NH_3}{|}}{CH}-COO^-$$

guanidinium
group

Histidine (His)

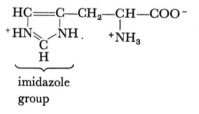

imidazole
group

Two α-amino acids are commonly mislabeled as basic when in fact they are neutral: *glutamine* and *asparagine*. They are the amides of glutamic and aspartic acids, respectively. Although they are polar, the amide groups neither protonate nor dissociate.

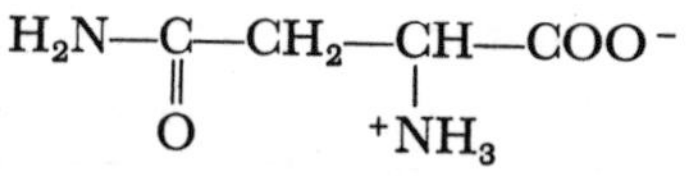

$$H_2N-\underset{\underset{O}{\|}}{C}-CH_2-\underset{\underset{^+NH_3}{|}}{CH}-COO^-$$

Asparagine (Asn)

$$H_2N-\underset{\underset{O}{\|}}{C}-CH_2-CH_2-\underset{\underset{^+NH_3}{|}}{CH}-COO^-$$

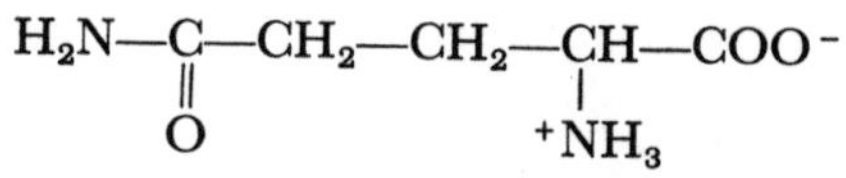

Glutamine (Gln)

In addition to their charge at neutral pH, amino acids may be classified according to whether or not they contain sulfur atoms, hydroxyl or aromatic groups, and branched or straight-chain hydrocarbons in their side chains (R groups). Each neutral amino acid may also be designated as *polar* or *nonpolar*.

Cysteine and *methionine* contain sulfur. The free sulfhydryl group of cysteine makes it a polar molecule. The —SH groups of cysteine can bind to one another to form the disulfide bridges that stabilize the structure of proteins. *Cystine* is a dimer of cysteine, in which two molecules of cysteine are joined via their sulfur atoms. Methionine, as the name implies, has a methylated thiol group (sulfur atom), which is nonpolar.

Methionine (Met)

$$CH_3—S—CH_2—CH_2—\underset{\underset{+NH_3}{|}}{CH}—COO^-$$

Cysteine (Cys)

$$HS—CH_2—\underset{\underset{+NH_3}{|}}{CH}—COO^-$$

sulfhydryl group

Cystine

$$\begin{array}{l} \underset{\underset{\underset{\underset{CH_2—\underset{+NH_3}{|}\ {CH}—COO^-}{|}}{S}}{S}}{CH_2}—\underset{\underset{+NH_3}{|}}{CH}—COO^- \end{array}$$

Three α-amino acids contain aromatic groups: *phenylalanine*, *tyrosine*, and *tryptophan*. Phenylalanine consists of a phenyl ring bound to the methyl group of alanine (which is shown subsequently). Tryptophan contains an indole group, which consists of a phenyl ring fused to a five-membered, nitrogen-containing ring. Both phenylalanine and tryptophan are nonpolar. The hydroxyl group of tyrosine, or *p*-hydroxyphenylalanine, renders it polar.

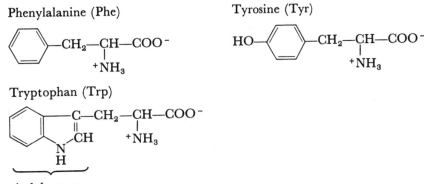

Phenylalanine (Phe)

Tyrosine (Tyr)

Tryptophan (Trp)

indole group

Three α-amino acids have branched hydrocarbon chains: *leucine*, *isoleucine*, and *valine*. Since their side chains are purely hydrocarbons, they are therefore nonpolar.

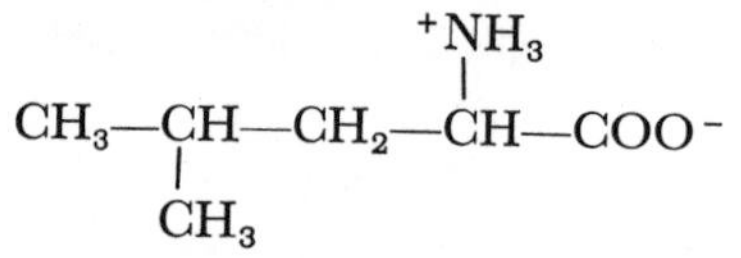

$$\overset{\displaystyle \overset{+NH_3}{|}}{CH_3-CH-CH_2-CH-COO^-} \atop \underset{CH_3}{|}$$

Leucine (Leu)

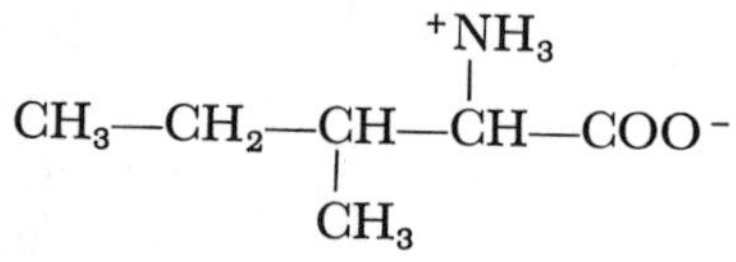

$$\overset{\displaystyle \overset{+NH_3}{|}}{CH_3-CH_2-CH-CH-COO^-} \atop \underset{CH_3}{|}$$

Isoleucine (Ile)

$$\overset{\displaystyle \overset{+NH_3}{|}}{CH_3-CH-CH-COO^-} \atop \underset{CH_3}{|}$$

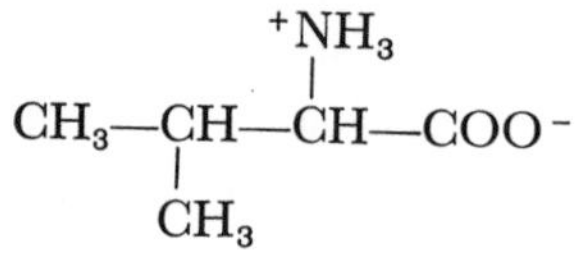

Valine (Val)

The R groups of *alanine* and *glycine* are a methyl group and a hydrogen atom, respectively. Glycine, the simplest of the amino acids, is highly polar, since it is dominated by its charged carboxyl and amino groups. The methyl side chain of alanine tends to make it nonpolar.

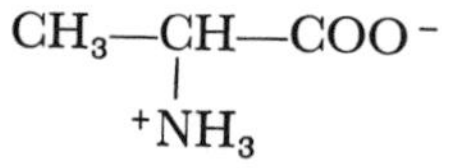

$$CH_3-\underset{\underset{+NH_3}{|}}{CH}-COO^-$$

Alanine (Ala)

$$\underset{\underset{+NH_3}{|}}{CH_2}-COO^-$$

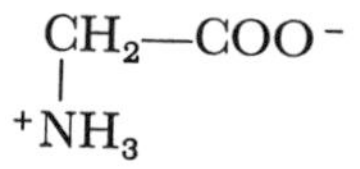

Glycine (Gly)

The R groups of *serine* and *threonine* contain hydroxyl groups, like tyrosine, which render them polar.

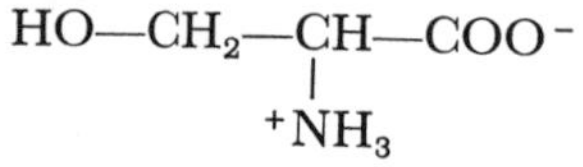

$$HO-CH_2-\underset{\underset{+NH_3}{|}}{CH}-COO^-$$

Serine (Ser)

$$CH_3-\underset{\underset{OH}{|}}{CH}-\underset{\underset{+NH_3}{|}}{CH}-COO^-$$

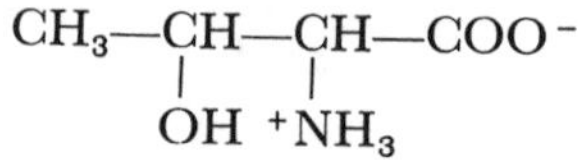

Threonine (Thr)

One so-called amino acid is actually an "imino" acid. The imino nitrogen in proline's five-membered ring binds to two carbon atoms. Because of the rigidity of this ring, proline residues will kink a chain of amino acids. Proline is nonpolar.

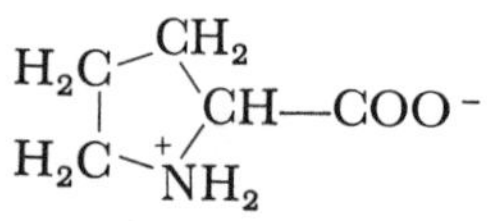

$$\begin{array}{c} H_2C-CH_2 \\ |\quad\quad\backslash \\ \quad\quad CH-COO^- \\ |\quad\quad / \\ H_2C-\overset{+}{N}H_2 \end{array}$$

Proline (Pro)

Ninhydrin reacts with the free α-amino groups of amino acids and proteins to produce a purple color. The ninhydrin reaction can be used to estimate the quantity of amino acid present in a sample. The quantitation of individual amino acids involves their separation by chromatographic techniques.

Describe each amino acid below in terms of its sulfur content, aromatic and hydroxyl groups, and branched or straight hydrocarbon chains. Designate each as acidic, neutral, or basic, and classify the neutral amino acids as polar or nonpolar.

7. $^+H_3N—CH_2—CH_2—CH_2—CH_2—CH—COO^-$
   $\overset{|}{{}^+NH_3}$

8. Phenylalanine

9. $CH_3—S—CH_2—CH_2—CH—COO^-$
   $\overset{|}{{}^+NH_3}$

10. Glycine

11. $CH_3—CH—CH—COO^-$
    $\overset{|}{CH_3}\ \overset{|}{{}^+NH_3}$

12. Glutamine

13. 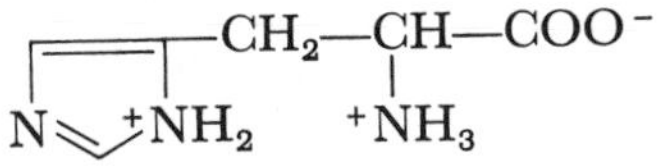 $\text{[imidazole ring]}—CH_2—CH—COO^-$
    $N{\nwarrow}\ {}^+NH_2\quad \overset{|}{{}^+NH_3}$

14. Proline

15. $^-OOC—CH_2—CH—COO^-$
    $\overset{|}{{}^+NH_3}$

16. Asparagine

17. $CH_3—CH—CH—COO^-$
    $\overset{|}{OH}\ \overset{|}{{}^+NH_3}$

18. Arginine

ANSWERS

1. Yes. This structure represents the simplest and smallest amino acid: glycine. Its R group is a hydrogen atom.

2. No. This molecule has a carboxyl group but lacks an amino group. It is a three-carbon fatty acid, propionic acid.

3. Yes. It has both an α-carboxyl and α-amino group. Its R group is an aromatic alcohol, which identifies this as tyrosine.

4. Although this molecule as a whole is not an amino acid, it does contain an amino acid. Look again carefully and find the amino acid:

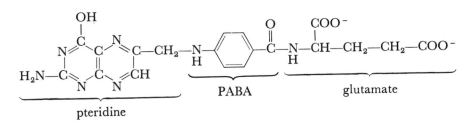

This molecule is the B vitamin, pteroylglutamic acid, one form of folic acid. The α-amino group of glutamic acid is linked through a peptide bond to *p*-aminobenzoic acid (PABA).

5. Did this β-amino acid fool you? Notice that the amino group binds to the β-carbon atom; hence, it is a β-amino acid, not an α-amino one.
6. This is a γ rather than an α-amino acid. It is γ-aminobutyric acid (GABA), an inhibitory neurotransmitter.
7. Basic (note the ε-amino group of lysine), straight-chain.
8. Neutral, aromatic, nonpolar.
9. Neutral, sulfur-containing (you should be able to identify this as methionine because the sulfur is methylated), nonpolar.
10. Neutral, polar.
11. Neutral, branched-chain, nonpolar.
12. Neutral (the amide does not protonate or dissociate), polar.
13. Basic (the imidazole group will have a positive charge; you should be able to identify this as histidine).
14. Neutral, nonpolar imino acid.
15. Acidic.
16. Neutral, polar.
17. The hydroxyl-containing group, like that of alcohols, is polar but does not dissociate; hence, it is a neutral amino acid.
18. Basic (the guanidine group of arginine has a positive charge at pH 7.0).

REFERENCES

Barker, R. *Organic Chemistry of Biological Compounds*. Englewood Cliffs, N.J.: Prentice-Hall, 1971. Pp. 54–76.

Lehninger, A. L. *Biochemistry: The Molecular Basis of Cell Structure and Function* (2nd ed.). New York: Worth, 1975. Pp. 71–76.

White, A., Handler, P., and Smith, E. L. *Principles of Biochemistry* (5th ed.). New York: McGraw-Hill, 1973. Pp. 89–96.

2 Acids, Bases, and Buffers

This may be one of the most challenging chapters in this book, and you should plan to spend at least fifteen hours to master this area. Never skip the problems; they are essential to your learning.

In biochemistry the most workable definitions of acids and bases are those of Brønsted, who defined an *acid* as a proton donor and a *base* as a proton acceptor. For each acid and each base, there is its *conjugate base* and *conjugate acid*, respectively, from which it differs by the proton lost or gained.

THE HENDERSON-HASSELBALCH EQUATION

Let RH represent an acid and R^- its conjugate base. Its dissociation may be represented by:

$$RH \rightleftharpoons H^+ + R^-$$

The ionization, or dissociation, constant of this acid, K_a, is defined by the equilibrium expression:

$$K_a = \frac{[H^+][R^-]}{[RH]} \tag{2-1}$$

where the square brackets indicate the molar concentrations of the substances. By rearrangement and substitution utilizing the definitions $pH = -\log[H^+]$ and $pK_a = -\log K_a$, we get:

$$[H^+] = \frac{K_a[RH]}{[R^-]}$$

$$-\log[H^+] = -\log K_a + \log \frac{[R^-]}{[RH]}$$

$$pH = pK_a + \log \frac{[R^-]}{[RH]} \tag{2-2}$$

Equation 2-2 is the *Henderson-Hasselbalch equation*, which may also be written:

$$pH = pK_a + \log \frac{[\text{proton acceptor}]}{[\text{proton donor}]}$$

Problem 1

Calculate the hydrogen ion concentration, $[H^+]$, in:

A. plasma at pH 7.4
B. gastric juice at pH 2.7

Problem 2

Calculate the pH of:

A. 0.001 M HCl
B. 0.20 M NaOH
C. 0.05 M CH_3COOH ($K_a = 1.86 \times 10^{-5}$)

Using the Henderson-Hasselbalch equation, compute the $[R^-]/[RH]$ ratio when:

A. $pH = pK_a$ D. $pH = pK_a - 1$
B. $pH = pK_a + 1$ E. $pH = pK_a - 2$
C. $pH = pK_a + 2$

If the pH differs by two or more units from the pK_a, you will not, for most practical purposes, need to use the Henderson-Hasselbalch equation to calculate the concentrations of the components of the dissociation reaction, because 99% or more of the substance will exist as the proton acceptor (if $pH \geq pK_a + 2$) or the proton donor (if $pH \leq pK_a - 2$).

The meaning of the Henderson-Hasselbalch equation can be easily conceptualized. If the pH drops below the pK_a, the conjugate base (R^-) is protonated to the acid (RH); hence, the ratio $[R^-]/[RH]$ falls below one. On the other hand, if the pH rises above the pK_a, the acid (RH) liberates its proton and the $[R^-]/[RH]$ ratio rises above one.

TITRATION AND BUFFERS

Titration is the incremental addition of a strong acid or base to a solution while measuring its pH up to the point, say, of neutralization. After the desired pH is reached, one calculates the moles of acid or base added, and from that figure, one determines the quantity of titratable acid or base in the solution.

The results of titration demonstrate whether or not the substance in solution is acting as a *buffer*, that is, a compound that changes pH relatively slowly in response to the addition of strong acid or base. Most buffers exhibit their buffering action only within a narrow pH range.

An important application of titration to medicine is in renal physiology. The *titratable acidity* of urine is defined as the number of millimoles of NaOH required to titrate one liter of urine up to physiologic pH (7.4). The principal titratable acid found in the urine is phosphate, which exists in three different forms and has two pK_a values:

$$H_3PO_4 \xrightleftharpoons[]{pK_{a_1} = 2.1} H^+ + H_2PO_4^- \xrightleftharpoons[]{pK_{a_2} = 7.2} H^+ + HPO_4^{-2}$$

Since the pH of urine never drops below 4.5, there is virtually no H_3PO_4 in urine, because $pK_{a_1} = 2.1$ (recall the value of $[R^-]/[RH]$ when $pH \geq pK_a + 2$).

Your patient, a Nobel prize-winning biochemist, asks you to calculate the ratio $[HPO_4^{-2}]/[H_2PO_4^-]$ in blood at pH 7.4 and urine at pH 7.2, 6.2, and 5.5. You mutter "Bullfeathers!" under your breath as you draw your pocket calculator and get to work.

The next morning the biochemist verifies your calculations. Now he wants you to compute the titratable acidity of the phosphate in urine at pH 7.2, 6.2, and 5.5, assuming the phosphate concentration is 0.01 M and ignoring other urinary buffers. Hearing you stammer a series of feeble excuses, he volunteers to help you. "For pH 7.2, you determined (see Problem 4) that $[HPO_4^{-2}]/[H_2PO_4^-] = 1.0$. Therefore, a 0.01 M (or 10 mM) phosphate solution at pH 7.2 will exist as 5.0 mM HPO_4^{-2} and 5.0 mM $H_2PO_4^-$. Now for pH 7.4, you determined (see Problem 4) that $[HPO_4^{-2}]/[H_2PO_4^-] = 1.6$. Hence, at pH 7.4, this solution will exist as 6.2 mM HPO_4^{-2}, since $(1.6/2.6) \times 10$ mM $=$ 6.2 mM, and as 3.8 mM $H_2PO_4^-$, since $(1/2.6) \times 10$ mM $= 3.8$ mM. In titrating 1 liter of this solution from pH 7.2 to pH 7.4, 1.2 mmoles of $H_2PO_4^-$ are converted to HPO_4^{-2} (5.0 mmoles − 3.8 mmoles = 1.2 mmoles $H_2PO_4^-$ converted). This liberates 1.2 mmoles of hydrogen ion:

$$1.2\ H_2PO_4^- \longrightarrow 1.2\ HPO_4^{-2} + 1.2\ H^+$$

which must be neutralized by NaOH:

$$1.2\ H^+ + 1.2\ NaOH \longrightarrow 1.2\ Na^+ + 1.2\ H_2O$$

In other words, 1.2 mmoles of NaOH are required to titrate 1 liter of this urine solution from pH 7.2 to physiologic pH. Thus, the titratable acidity of one liter of urine containing 10 mM phosphate at pH 7.2 is 1.2 mmoles. Now follow this example for the cases of pH 6.2 and pH 5.5."

Amino Acids as Buffers

Proteins function as one of the most important buffer systems in blood and tissues, and their buffering ability derives from the dissociable groups on their constituent amino acids.

Glycine, the simplest amino acid, has two dissociable groups, the α-amino and α-carboxyl groups:

$$^+H_3N{-}CH_2{-}COOH \overset{pK_{a_1} = 2.3}{\rightleftharpoons}$$
$$^+H_3N{-}CH_2{-}COO^- \overset{pK_{a_2} = 9.6}{\rightleftharpoons} H_2N{-}CH_2{-}COO^-$$

Glycine can therefore exist in three forms, depending on the pH:

1. Completely protonated ($^+H_3N{-}$, $-COOH$). Net charge of positive one.
2. Protonated α-amino group with unprotonated α-carboxyl group ($^+H_3N{-}$, $-COO^-$). This is the *isoelectric* species of glycine because it has zero net charge.
3. Completely unprotonated ($H_2N{-}$, $-COO^-$). Net charge of negative one.

To determine the form and net charge of glycine at pH 3.0, for example, consider each dissociable group separately. Since pK_{a_1} for the α-carboxyl group is 2.3, this group will exist mainly as COO^-. More than about 10%, however, will be in the $COOH$ form, because $pH - pK_a < 1.0$. Applying the Henderson-Hasselbalch equation,

$$3.0 = 2.3 + \log[COO^-]/[COOH]$$
$$[COO^-]/[COOH] = 5$$

9

Therefore, $\frac{5}{6}$ of the carboxyl groups will exist as COO^-, while $\frac{1}{6}$ will be $COOH$. The net charge due to the carboxyl groups will be $-\frac{5}{6}$.

Turning to the α-amino group, its pK_{a_2} of 9.6 is 6.6 units above pH 3. You do not need to use the Henderson-Hasselbalch equation here, because virtually 100% of this group will exist as $^+NH_3$ with a charge of $+1$. Thus, the net charge on glycine at pH 3.0 is $+\frac{1}{6}$.

The net charge on an amino acid or protein determines its mobility in *electrophoresis*, a technique that is used to separate compounds of differing charge by applying an electrical potential across a medium, such as paper or starch gel, that contains a solution of buffer and the sample compounds. Since glycine at pH 3.0 has a charge of $+\frac{1}{6}$, it will move toward the negative pole during electrophoresis at this pH.

Because its imidazole group has a pK_a of 6.0, histidine is the only amino acid with significant buffering capacity in the physiologic range, i.e., between pH 6 and pH 8. The pK_a of cysteine's sulfhydryl group is 8.3, which is closer to the normal blood pH of 7.44 than the pK_a of the imidazole group, but this SH group is often tied up in disulfide linkages (such as in the dimer cystine) and is therefore unable to act as a buffer.

Let us next consider the buffering properties of histidine:

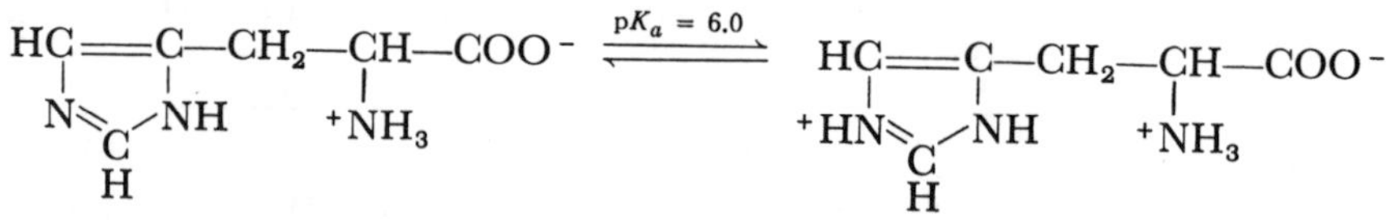

At pH 7.4 the α-carboxyl group ($pK_a = 1.8$) will exist as COO^-, while the α-amino group ($pK_a = 9.0$) will exist almost entirely as $^+NH_3$. For the imidazole group,

$$7.4 = 6.0 + \log \frac{[N\diagup]}{[^+HN\diagup]}$$

$$\frac{[N\diagup]}{[^+HN\diagup]} = 25$$

Hence, at this pH, the charge due to this group is $+\frac{1}{26}$. Since the -1 charge on the α-carboxyl group balances the $+1$ charge on the α-amino group, the net charge on histidine at pH 7.4 is $+\frac{1}{26}$.

Problem 6

Hemoglobin A (HbA), the predominant adult hemoglobin, contains 5% histidine by weight. How many millimoles of histidine are in one liter of blood containing 14 g hemoglobin per 100 ml blood, if the histidine content of other blood proteins is ignored? The molecular weight (MW) of His is 156 g/mole.

In titrating hemoglobin starting at pH 7.4, we need consider only the imidazole groups of histidine until the pH rises above 7.6 or falls below 5.5, because no other dissociable R groups on the hemoglobin molecule have pK_a values near 7.4. If we titrate a one liter solution of 14 g HbA per 100 ml starting at pH 7.4, we are in fact titrating a 45 mM solution of the imidazole group of histidine (see Problem 6). At pH 7.4 the protonated form of the imidazole group accounts for 1.7 millimoles ($\frac{1}{26}$ of 45 mM) and the unprotonated form for 43.3 millimoles per liter ($\frac{25}{26}$ of 45 mM). If 2.0 mmoles HCl are added, the following buffering reaction occurs:

$$2 \text{ mmoles HCl} + 2 \text{ mmoles N} \diagup \longrightarrow 2 \text{ mmoles } {}^+\text{HN} \diagup + 2 \text{ mmoles Cl}^-$$

Thus, there will now be 3.7 millimoles protonated and 41.3 millimoles unprotonated imidazole groups. Applying the Henderson-Hasselbalch equation,

$$pH = 6.0 + \log\left(\frac{41.3 \text{ mM}}{3.7 \text{ mM}}\right)$$

$$pH = 6.0 + \log(11.2) = 7.05$$

Problem 7

Using the method just given, complete the titration chart below for one liter of hemoglobin solution (14 g/100 ml) starting at pH 7.4:

Millimoles HCl added	$\dfrac{[\text{N}\diagup]}{[{}^+\text{HN}\diagup]}$	pH
2	11.3	7.05
4	. . .	. . .
8	. . .	. . .
14	. . .	. . .
20.8	. . .	. . .

Problem 8

Using the pK_a values listed below, identify amino acids A, B, C, and D from their titration curves in Figure 2-1.

Amino acid	pK_{a_1}	pK_{a_2}	pK_{a_3}
Asp	1.9	3.6	9.6
His	1.8	6.0	9.2
Lys	2.2	9.0	10.5
Thr	2.6	10.4	—

Figure 2-1

Titration curves for amino acids in Problem 8.

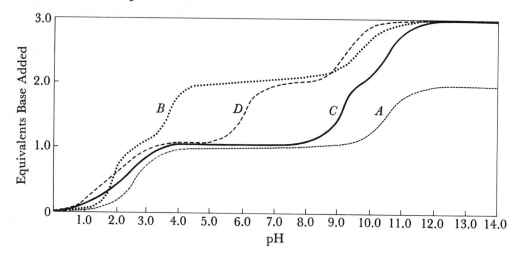

Isoelectric pH

The *isoelectric pH*, or pH_I, is the pH at which an amino acid has a net charge of zero. For an amino acid with two dissociable groups, the pH_I is the average of the two respective pK_a values. For an amino acid with three dissociable groups, one must average the two pK_a values on either side of the isoelectric species to obtain the pH_I. Thus, for cysteine,

$$HS, {}^+NH_3, COOH \xrightleftharpoons{pK_{a_1} = 1.7} HS, {}^+NH_3, COO^- \xrightleftharpoons{pK_{a_2} = 8.3}$$

charge = +1   isoelectric,
charge = 0

$$ {}^-S, {}^+NH_3, COO^- \xrightleftharpoons{pK_{a_3} = 10.8} {}^-S, NH_2, COO^-$$

charge = −1   charge = −2

Thus, the pH_I of cysteine is the average of $pK_{a_1}(1.7)$ and $pK_{a_2}(8.3)$, which equals 5.0.

Problem 9

Calculate the pH_I for the amino acids listed in the table under Problem 8.

Ion Trapping

One of the principal means of excreting acids into the urine is the renal tubular secretion of $NH_4{}^+$ (ammonium). Once inside the lumen of the tubules, ammonium cannot be reabsorbed unless it is converted to ammonia (NH_3). In other words, the $NH_4{}^+$ cation becomes trapped within the lumen. Since the pK_a of ammonium is 9.3, within the normal urinary pH range of 4.5 to 7 virtually all the NH_3 secreted by the tubules is trapped as $NH_4{}^+$.

Another application of ion trapping to medicine is to augment the excretion of barbiturates after an overdose by administering agents to render the urine more alkaline.

The Bicarbonate Buffer

Bicarbonate (HCO_3^-) provides the most important physiologic buffer system for several reasons:

1. Bicarbonate is present in high concentration in plasma; e.g., in humans, its concentration is 25 mEq/liter at pH 7.4 when $P_{CO_2} = 40$ mm Hg.
2. The enzyme carbonic anhydrase in erythrocytes (red blood cells) and renal tubules allows the rapid interconversion of CO_2 and carbonic acid (H_2CO_3).
3. The lungs regulate the partial pressure of CO_2 (P_{CO_2}) in the blood on a minute-by-minute basis.
4. The kidneys regulate urinary bicarbonate excretion and can alter the plasma HCO_3^- concentration slowly, over the course of hours.
5. Hemoglobin is positioned side-by-side with carbonic anhydrase in erythrocytes and assists the buffering action of bicarbonate.

Let us first consider the bicarbonate buffer system in vitro, unassisted by the lungs, kidneys, carbonic anhydrase, or hemoglobin.

$$CO_2 + H_2O \rightleftharpoons H_2CO_3 \underset{}{\overset{pK_a = 6.1}{\rightleftharpoons}} H^+ + HCO_3^-$$

In vitro, the hydration of CO_2 to form carbonic acid, H_2CO_3, and the reverse reaction occur very slowly. As a buffer at pH 7.4, this system compares in strength to the histidine buffer. If we add HCl to this buffer, the rapid net reaction will be:

$$H^+ + HCO_3^- \longrightarrow H_2CO_3$$

As the pH is raised, carbonic acid will dehydrate to yield CO_2, which can escape into the surrounding air.

Adding the enzyme carbonic anhydrase in vitro improves the buffer strength of the bicarbonate buffer. If HCl is added to such a system, the rapid net reaction will be:

$$H^+ + HCO_3^- \longrightarrow H_2CO_3 \overset{\text{Carbonic anhydrase}}{\longrightarrow} CO_2 + H_2O$$

Applying the Henderson-Hasselbalch equation, we get:

$$pH = 6.1 + \log \frac{[HCO_3^-]}{[H_2CO_3]}$$

In the presence of carbonic anhydrase, the H_2CO_3 concentration will be so small that it cannot be accurately measured. Instead, we may substitute the CO_2 concentration (mmoles/liter):

$$pH = 6.1 + \log \frac{[HCO_3^-]}{[CO_2]}$$

In practice, we measure the partial pressure of carbon dioxide (P_{CO_2} in mm Hg), rather than measuring the CO_2 concentration directly:

$$[CO_2] \text{ (mmoles/liter)} = 0.03 P_{CO_2} \text{ (mm Hg)}$$

13

Therefore,

$$\text{pH} = 6.1 + \log \frac{[\text{HCO}_3{}^-]}{0.03 P\text{co}_2}$$

Adding HCl to this system consumes $\text{HCO}_3{}^-$, thereby lowering the $\text{HCO}_3{}^-$ concentration and raising the $P\text{co}_2$. Because it is volatile, some of this CO_2 will enter the air and lower the amount of dissolved CO_2.

Now let us consider the in vivo bicarbonate buffer system that utilizes the lungs and kidneys, but, hypothetically, lacks hemoglobin and other buffers. If we add HCl, we will rapidly lower the $\text{HCO}_3{}^-$ concentration and raise the $P\text{co}_2$. Immediately, the lungs will compensate by increasing the ventilation through augmenting the tidal volume, the respiratory rate, or both, to boost the CO_2 loss from the lungs and lower blood $P\text{co}_2$. Over a period of several hours, the kidneys will increase their tubular reabsorption of bicarbonate to restore slowly the serum $\text{HCO}_3{}^-$ level. Thus, the renal control of urinary bicarbonate excretion along with the pulmonary control of CO_2 exchange make this in vivo buffer system much more powerful in the long run than the other physiologic buffer systems.

The Hemoglobin Buffer

To handle sudden additions of acid or base, another buffer system is called into play which is more powerful *in the short run* than the bicarbonate system; this is the histidine buffer of hemoglobin. (It is a magnificent plan to have both the carbonic anhydrase for the bicarbonate buffer and hemoglobin housed together inside the erythrocytes.)

As you calculated in Problem 6, the concentration of hemoglobin-bound histidine in whole blood is 45 mM when there are 14 g hemoglobin per 100 ml of blood. To lower the pH of this hemoglobin buffer from 7.4 to 7.0, one would have to add almost 2 millimoles HCl per liter of blood.

The protonation of the imidazole groups of hemoglobin is coupled to the deoxygenation of hemoglobin:

$$\text{HbO}_2 + \text{H}^+ \rightleftharpoons \text{HHb}^+ + \text{O}_2$$

Thus, the oxygen-binding properties of hemoglobin are intimately related to its buffer properties. Figure 2-2 shows the relation of $P\text{o}_2$ and pH to HbO_2 saturation at 37°C and $P\text{co}_2 = 40$ mm Hg. The sigmoidal shape of the O_2 dissociation curve (which will be explained in Chapter 3) indicates that the initial binding of O_2 to the Fe^{+2} of the Hb tetramer enhances subsequent binding of more O_2 molecules.

The O_2 dissociation curve is shifted to the right by acidosis (arterial pH < 7.4), hypercapnia (elevated $P\text{co}_2$), fever, or increased 2,3-diphosphoglycerate (2,3-DPG) levels within the erythrocytes. Note in Figure 2-2 that at pH 7.2, the rightward shift of the curve indicates reduced affinity of Hb for O_2.

This curve is shifted to the left by alkalosis (arterial pH > 7.46), hypocapnia, and hypothermia (low body temperature). In Figure 2-2, the leftward shift at pH 7.6 indicates increased affinity of Hb for O_2.

The P_{50} of hemoglobin is defined as the $P\text{o}_2$ at which Hb is 50% saturated.

Figure 2-2   *Oxygen dissociation curves of hemoglobin.*

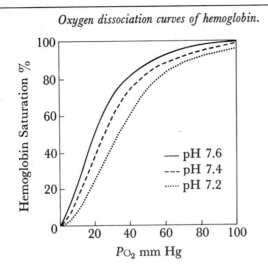

Figure 2-2 plot: Hemoglobin Saturation % (y-axis, 0 to 100) vs P_{O_2} mm Hg (x-axis, 20 to 100).

Legend:
— pH 7.6
--- pH 7.4
····· pH 7.2

Problem 10

Read the P_{50} of Hb at pH 7.2, 7.4, and 7.6 from Figure 2-2. How does a rightward shift influence the P_{50}?

At extreme P_{O_2} values (below 10 mm Hg or above 85 mm Hg), any shift in the curve has little effect on O_2 saturation. Thus, the P_{O_2} of 90 to 100 mm Hg in the pulmonary capillaries will saturate Hb, despite, say, a rightward shift of the dissociation curve.

At a P_{O_2} of 40 mm Hg in tissue capillaries, hemoglobin at pH 7.2 and $P_{CO_2} = 40$ mm Hg will be 55% saturated. In other words, if the Hb is saturated when it leaves the pulmonary circulation, it will unload 45% of its O_2 content in the tissues. Hemoglobin at pH 7.4 will unload only 25% of its O_2 content under similar conditions, because at a P_{O_2} of 40 mm Hg it is 75% saturated. Thus, a rightward shift improves the tissue O_2 unloading, provided that the lung can saturate this hemoglobin.

As it traverses the pulmonary capillaries, deoxyhemoglobin takes up O_2 and simultaneously releases the H^+ bound to its imidazole group. Two ways to represent this reaction are:

$$HbH^+ + O_2 \rightleftharpoons HbO_2 + H^+$$

$$(Fe^{+2}, {}^+HN{\scriptstyle\lessgtr}) + O_2 \rightleftharpoons (Fe^{+2}{-}O_2, N{\scriptstyle\lessgtr}) + H^+$$

The H^+ liberated inside the red blood cell immediately combines with HCO_3^- to form H_2CO_3. This carbonic acid is then dehydrated by carbonic anhydrase to yield $CO_2 + H_2O$. The CO_2 diffuses out of the red blood cell, and much of it diffuses through the alveolocapillary membrane into the alveolar airspace.

In the tissue capillaries, on the other hand, HbO_2 unloads much of its O_2 because of its reduced affinity for O_2 at the lower P_{O_2} levels (40 to 50 mm Hg) found there. This deoxygenation, which is accompanied by protonation of the imidazole group of histidine, is the opposite reaction of the oxygenation and proton liberation shown above.

By accepting protons, Hb helps buffer the blood against pH changes due to the metabolic acids produced by the tissues, while it simultaneously supplies the tissues with O_2. As shown above, these protons, which are bound to histidine, are unloaded in the pulmonary capillaries when the Hb is oxygenated.

Of less significance is the addition of CO_2 to the N-terminal amino groups of the α and β chains of hemoglobin to form carbamino compounds. Such compounds carry only 5% of the total CO_2 content of the blood.

Problem 11

Carbon monoxide (CO) is toxic by virtue of its 200-fold greater affinity for Hb compared to that of O_2 for Hb. What three measures will benefit a patient with carbon monoxide poisoning?

Problem 12

Determine the O_2 content of Hb under the following conditions: pH = 7.6, Po_2 = 100 mm Hg, [Hb] = 15 g/100 ml (see Figure 2-2 to determine Hb saturation). Use the formula:

$$\frac{ml\ O_2}{100\ ml\ blood} = \left(\frac{g\ Hb}{100\ ml\ blood}\right) \times \left(\frac{Hb\ saturation\ \%}{100}\right) \times \left(\frac{1.34\ ml\ O_2}{g\ Hb}\right)$$

What fraction of its O_2 content will this Hb unload at a tissue Po_2 of 40 mm Hg?

Problem 13

What is the resultant pH if 10 ml of 2.0 M NaOH is added to 1.0 liter of pH 11.0 buffer consisting of 0.2 M asparagine (pK_{a_1} = 2.02, pK_{a_2} = 8.80) and 0.1 M lysine (pK_{a_1} = 2.18, pK_{a_2} = 8.95, pK_{a_3} = 10.5)? (This is an easy problem once you decide what will and what will not react with the NaOH.)

Problem 14

When the pH is below the pH_I of an amino acid, what is its net charge: negative, zero, or positive?

Problem 15

A sample of aspartic acid is titrated from pH 1.0 to pH 6.5 by the addition of 3.0 mmoles KOH. Without performing any written calculations you should be able to predict the approximate number of millimoles of aspartic acid in the sample (pK_{a_1} = 1.88, pK_{a_2} = 3.65, pK_{a_3} = 9.60).

A. 1.0
B. 1.5

C. 2.0
D. 3.0

ACID-BASE IMBALANCES

Acidosis in adults is defined as an arterial pH below 7.40, while *alkalosis* is defined as an arterial pH above 7.46.

Primary respiratory acidosis exists when there is acidosis plus hypercapnia (Pco_2 > 45 mm Hg). The cause may be either acute pulmonary insufficiency (e.g., asthma, pneumonia, pulmonary embolism) or chronic pulmonary insufficiency (as in chronic obstructive pulmonary disease).

Primary respiratory alkalosis occurs when there is alkalosis plus hypocapnia (Pco_2 < 35 mm Hg). Causes include hyperventilation due to severe anxiety or increased minute-ventilation in patients on respirators. Increased renal excretion of HCO_3^- may partially correct the alkalosis; the condition is then termed *primary respiratory alkalosis* with metabolic compensation.

Primary metabolic acidosis exists when a low serum level of bicarbonate ($[HCO_3^-] < 22$ mEq/liter) accompanies acidosis. The anion gap, which is defined as $[Na^+] - [HCO_3^-] - [Cl^-]$, allows a distinction to be made between two categories of primary metabolic acidosis. A *normal anion gap* (8–16 mEq/liter) may be found in cases of HCO_3^- wasting due to various causes (diarrhea, the use of carbonic anhydrase inhibitors, pancreatic fistulas, or proximal renal tubular acidosis) and in metabolic acidosis as a result of NH_4Cl administration. An *increased anoin gap* (≥ 16 mEq/liter) is found when an anion other than HCO_3^- or Cl^- is causing the acidosis; examples include lactic acidosis, keto-acidosis in diabetes or in alcoholism, and aspirin poisoning. In most cases of primary metabolic acidosis, respiratory compensation takes place, tending to return the pH toward normal. In fact, hyperventilation is a valuable diagnostic clue; it occurs with all types of acute primary metabolic acidosis.

Primary metabolic alkalosis occurs when a high serum bicarbonate level ($[HCO_3^-] > 29$ mEq/liter) accompanies alkalosis. Causes include a loss of gastric HCl without pancreatic HCO_3^- loss (e.g., vomiting with pyloric obstruction) and the administration of diuretics. Respiratory compensation does not occur to a significant extent.

Regardless of the type of acidosis or alkalosis, the Henderson-Hasselbalch equation can always be used to calculate the pH, P_{CO_2}, or the HCO_3^- level, as long as the values of two of these variables are known.

Problem 16

A patient arrives at the emergency room in a coma due to a drug overdose. The arterial P_{O_2} is 50 mm Hg (normal 70–100), the arterial P_{CO_2} is 60 mm Hg, and the HCO_3^- level is 34 mEq/liter. Calculate his arterial pH using the Henderson-Hasselbalch equation. How would you classify his acid-base status?

Problem 17

As he dashes to the toilet, your new patient tells you that he has had persistent diarrhea for 7 days. You collect a stool specimen, which, on microscopic examination, is found to be teaming with the trophozoites of *Entamoeba histolytica*. His data are $P_{O_2} = 90$ mm Hg, $P_{CO_2} = 42$ mm Hg, $[HCO_3^-] = 18$ mEq/liter, $[Na^+] = 135$ mEq/liter, and $[Cl^-] = 107$ mEq/liter. Calculate his arterial pH and anion gap, and classify his acid-base disturbance.

Problem 18

Your patient, who is receiving oxygen therapy because of chronic obstructive lung disease, has an arterial P_{O_2} of 50 mm Hg, P_{CO_2} of 30 mm Hg, and pH of 7.50. Calculate his HCO_3^- level, and classify his acid-base status.

Problem 19

A patient who is receiving diuretics for congestive heart failure has an arterial pH of 7.50, HCO_3^- level of 31 mEq/liter, and a low serum K^+ level of 2.8 mEq/liter. What is his P_{CO_2}? What type of acid-base disturbance does he have?

Explain why the serum phosphate buffer is a weaker buffer than the hemoglobin system, despite the fact that the pK_{a_2} of phosphate, 7.2, is closer to physiologic pH than is the pK_a for the imidazole group of the histidine of Hb. The serum phosphate level is 0.3 to 0.4 mM.

ANSWERS

1. A. $\text{Log}[H^+] = 0.6 - 8$; $[H^+] = 4.0 \times 10^{-8}$ M
   B. $\text{Log}[H^+] = 0.3 - 3$; $[H^+] = 2.0 \times 10^{-3}$ M

2. A. Being a strong acid, HCl dissociates completely. Hence, $[H^+] = 1 \times 10^{-3}$ M, and pH = 3.0.
   B. As a strong base, NaOH dissociates completely. Thus, $[OH^-] = 0.20$ M. For water, we know that $[H^+] \times [OH^-] = 1 \times 10^{-14}$. Therefore, $[H^+] = 5 \times 10^{-14}$ M, and pH = 13.3.
   C. $CH_3COOH \rightleftharpoons H^+ + CH_3COO^-$

   $$K_a = \frac{[H^+][CH_3COO^-]}{[CH_3COOH]}$$

   Let $[CH_3COO^-] = x$. We can assume that $[H^+] = x$, because in a solution of a weak acid, such as acetic acid, virtually all the H^+ is derived from the dissociation of CH_3COOH to form equimolar amounts of H^+ and CH_3COO^-; the contribution due to the dissociation of water ($K_w = 1 \times 10^{-14}$) is insignificant in such circumstances. Therefore, $[CH_3COOH] = 0.05 - x$. Substituting into the equilibrium equation (Eq. 2-1), we get:

   $$1.86 \times 10^{-5} = \frac{x^2}{0.05 - x}$$

   Since this is a weak acid, the amount dissociated is small relative to the initial amount, i.e., $0.05 \gg x$, so we can simplify to:

   $$1.86 \times 10^{-5} = \frac{x^2}{0.05}$$

   $$x^2 = 0.930 \times 10^{-6}$$

   $$x = 0.965 \times 10^{-3}$$

   $$\text{pH} = 3.0$$

3. A. Let $[R^-]/[RH] = x$. When $pH = pK_a$, $\log x = 0$, and $x = 1$. Thus, at this pH, the group is 50% protonated and 50% unprotonated.
   B. $\text{Log } x = 1$; $x = 10$
   C. $\text{Log } x = 2$; $x = 100$
   D. $\text{Log } x = -1$; $x = 0.1$
   E. $\text{Log } x = -2$; $x = 0.01$
   Thus, for every one-unit difference between the pH and the pK_a, the ratio of base to acid changes by a factor of 10.

4. $\text{pH} = pK_a + \log \dfrac{[HPO_4^{-2}]}{[H_2PO_4^{-}]}$

For blood at pH 7.4,

$$7.4 = 7.2 + \log \frac{[HPO_4^{-2}]}{[H_2PO_4^-]}$$

$$\log \frac{[HPO_4^{-2}]}{[H_2PO_4^-]} = 0.2$$

$$\frac{[HPO_4^{-2}]}{[H_2PO_4^-]} = 1.6$$

For urine at pH 7.2,

$$\log \frac{[HPO_4^{-2}]}{[H_2PO_4^-]} = 0$$

$$\frac{[HPO_4^{-2}]}{[H_2PO_4^-]} = 1$$

For urine at pH 6.2,

$$\log \frac{[HPO_4^{-2}]}{[H_2PO_4^-]} = -1$$

$$\frac{[HPO_4^{-2}]}{[H_2PO_4^-]} = 0.1$$

For urine at pH 5.5,

$$\log \frac{[HPO_4^{-2}]}{[H_2PO_4^-]} = -1.7$$

$$\frac{[HPO_4^{-2}]}{[H_2PO_4^-]} = 0.02$$

5. *At pH 6.2,*

$$\frac{[HPO_4^{-2}]}{[H_2PO_4^-]} = 0.1$$

Hence, a 10 mM solution of phosphate at pH 6.2 will exist as 9.1 mM $H_2PO_4^-$ ($\frac{10}{11}$ of 10 mM) and 0.9 mM HPO_4^{-2} ($\frac{1}{11}$ of 10 mM). To titrate one liter of this solution from pH 6.2 to pH 7.4 would require 5.3 mmoles NaOH to lower $[H_2PO_4^-]$ from 9.1 mM to 3.8 mM:

5.3 mmoles $H_2PO_4^-$ + 5.3 mmoles NaOH $\longrightarrow$

5.3 mmoles HPO_4^{-2} + 5.3 mmoles Na^+ + 5.3 mmoles H_2O

Thus, the titratable acidity of one liter of urine containing 10 mM phosphate at pH 6.2 is 5.3 mmoles NaOH.

At pH 5.5,

$$\frac{[HPO_4^{-2}]}{[H_2PO_4^{-}]} = 0.02$$

Hence, a 10 mM solution of phosphate at pH 5.5 will exist as 9.8 mM $H_2PO_4^{-}$ (50/51 of 10 mM) and 0.2 mM HPO_4^{-2} (1/51 of 10 mM). To titrate one liter of solution from pH 5.5 to pH 7.4 would require 6.0 mmoles NaOH to lower $[H_2PO_4^{-}]$ from 9.8 mM to 3.8 mM. Thus, the titratable acidity of one liter of urine containing 10 mM phosphate at pH 5.5 is 6.0 mmoles NaOH. (Compare this result with that of the example in the text. Note that proportionally more NaOH is required to change the pH from 7.2 to 7.4 than to go from pH 5.5 to 7.4; this illustrates the buffering action of phosphate at pH values around the pK_a of 7.2.

6. There are 140 g Hb per liter of blood.

    g His/liter $= 0.05 \times 140$ g/liter $= 7.0$ g/liter

    moles His/liter $= (7.0 \text{ g})/(156 \text{ g/mole}) = 0.045$ M or 45 mmoles/liter

7. *After adding 4 mmoles HCl*, the imadazole group exists as 5.7 mM protonated form and 39.3 mM unprotonated. The ratio of these concentrations is 6.90:

    $$pH = 6.0 + \log\left(\frac{39.3 \text{ mM}}{5.7 \text{ mM}}\right)$$

    $$pH = 6.0 + \log(6.90) = 6.84$$

    After adding 8 mmoles HCl, the imadazole exists as 9.7 mM protonated and 35.3 mM unprotonated; the concentration ratio is 3.64:

    $$pH = 6.0 + \log(3.64) = 6.56$$

    After adding 14 mmoles HCl, the imidazole group exists as 15.7 mM protonated and 29.3 mM unprotonated; the concentration ratio is 1.87:

    $$pH = 6.0 + \log(1.87) = 6.27$$

    After adding 20.8 mmoles HCl, the imidazole group exists as 22.5 mM protonated and 22.5 mM unprotonated; the concentration ratio is 1.00:

    $$pH = 6.0 + \log(1.00) = 6.00$$

    Curve *D* (histidine) in Figure 2-1 (see Problem 8) shows that only a small amount of acid need be added to change a histidine buffer from pH 8.0 to 7.0; much more acid is needed to lower the pH further to 6.0. Any buffer resists pH changes best when the pH approaches its pK_a.

8. Curve *A* is Thr (pK_a 2.6, 10.4).
    Curve *B* is Asp (pK_a 1.9, 3.6, 9.6).
    Curve *C* is Lys (pK_a 2.2, 9.0, 10.5).
    Curve *D* is His (pK_a 1.8, 6.0, 9.2).

9. Aspartic acid has three dissociable groups; hence, one must draw all the species to determine which two pK_a values to average in calculating pH$_I$:

$^+$NH$_3$, COOH, COOH $\xrightleftharpoons{\text{p}K_{a_1} = 1.9}$ $^+$NH$_3$, COO$^-$, COOH $\xrightleftharpoons{\text{p}K_{a_2} = 3.6}$

charge = +1   isoelectric

$^+$NH$_3$, COO$^-$, COO$^-$ $\xrightleftharpoons{\text{p}K_{a_3} = 9.6}$ NH$_2$, COO$^-$, COO$^-$

charge = −1   charge = −2

Thus,

$$\text{pH}_I = \frac{1.9 + 3.6}{2} = 2.75$$

For histidine:

$^+$HN$\backslash\!\!\backslash$, $^+$NH$_3$, COOH $\xrightleftharpoons{\text{p}K_{a_1} = 1.8}$ $^+$HN$\backslash\!\!\backslash$, $^+$NH$_3$, COO$^-$ $\xrightleftharpoons{\text{p}K_{a_2} = 6.0}$

charge = +2   charge = +1

N$\backslash\!\!\backslash$, $^+$NH$_3$, COO$^-$ $\xrightleftharpoons{\text{p}K_{a_3} = 9.2}$ N$\backslash\!\!\backslash$, NH$_2$, COO$^-$

isoelectric   charge = −1

Hence,

$$\text{pH}_I = \frac{6.0 + 9.2}{2} = 7.6$$

For lysine:

$^+$NH$_3$, $^+$NH$_3$, COOH $\xrightleftharpoons{\text{p}K_{a_1} = 2.2}$ $^+$NH$_3$, $^+$NH$_3$, COO$^-$ $\xrightleftharpoons{\text{p}K_{a_2} = 9.0}$

charge = +2   charge = +1

$^+$NH$_3$, NH$_2$, COO$^-$ $\xrightleftharpoons{\text{p}K_{a_3} = 10.5}$ NH$_2$, NH$_2$, COO$^-$

isoelectric   charge = −1

$$\text{pH}_I = \frac{9.0 + 10.5}{2} = 9.75$$

Threonine has only two dissociable groups. Therefore,

$$\text{pH}_I = \frac{2.6 + 10.4}{2} = 6.5$$

10. P_{50} at pH 7.2 is 35 mm Hg.

      P_{50} at pH 7.4 is 27 mm Hg.

      P_{50} at pH 7.6 is 22 mm Hg.

11. *First*, remove the CO source to lower P_{CO}. *Second*, give supplemental O_2 and assisted ventilation, if needed, so that O_2 will have a better chance to compete for Hb binding. A recent breakthrough is the use of hyperbaric O_2, i.e., O_2 administered above atmospheric pressure. *Third*, perform exchange transfusion of red blood cells to remove HbCO complexes (this is rarely used clinically).

12.
$$\frac{ml\ O_2}{100\ ml\ blood} = 15\ g\ Hb/100\ ml \times \frac{100\%}{100} \times 1.34\ ml\ O_2/g\ Hb$$

$$= O_2\ content = 20.1\frac{ml\ O_2}{100\ ml\ blood}$$

At P_{O_2} = 40 mm Hg, the Hb saturation = 80%; thus, Hb will liberate 20% of its O_2 content at this pressure.

13. The only group that reacts significantly with NaOH is the ϵ-amino group of lysine, because the pK_a values of the other groups are more than two units from the pH. In one liter of this solution, there are 100 mmoles of the ϵ-amino group. At pH 11.0:

$$11.0 = 10.5 + \log \frac{[-NH_2]}{[-^+NH_3]}$$

$$\frac{[-NH_2]}{[-^+NH_3]} = 3.2$$

$$mmoles-NH_2 = \frac{3.2 \times 100}{4.2}\ mmoles = 76\ mmoles$$

After adding 20 mmoles of NaOH, there will be 96 mmoles—NH_2 and 4 mmoles—$^+NH_3$:

$$pH = 10.5 + \log(\tfrac{96}{4}) = 11.88$$

14. Below the pH_I an amino acid carries a positive charge, because its groups exist mainly in the protonated form.

15. The correct choice is B, 1.5 mmoles aspartic acid. Starting at pH 1.0, the KOH must titrate about 90% of the α-carboxyl group (pK_{a_1} 1.88) to reach pH 6.5, along with 100% of the β-carboxyl group:

$$HOOCCH_2CH(NH_3{}^+)COOH + 2OH^- \longrightarrow {}^-OOCCH_2CH(NH_3{}^+)COO^- + 2H_2O$$

16.
$$pH = 6.1 + \log\left(\frac{34\ mEq/liter}{0.03 \times 60\ mm\ Hg}\right) = 7.37$$

He has primary respiratory acidosis because he is acidotic and hypercapnic. His hypoxemia (low arterial P_{O_2}) is due to inadequate ventilation caused by the drug overdose. In addition, he has compensatory metabolic alkalosis as indicated by his high serum $HCO_3{}^-$ level.

17. Anion gap = $[Na^+] - [Cl^-] - [HCO_3^-] = 135 - 107 - 18 = 10$ mEq/liter. This is a normal anion gap.

$$pH = 6.1 + \log\left(\frac{18 \text{ mEq/liter}}{0.03 \times 42 \text{ mm Hg}}\right) = 7.26$$

This poor fellow with intestinal amebiasis has primary metabolic acidosis with no excessive anions other than HCO_3^- and Cl^- (i.e., unmeasured anions), as determined by a normal anion gap, and no compensatory respiratory alkalosis.

18. $$7.50 = 6.10 + \log\left(\frac{[HCO_3^-]}{0.03 \times 30 \text{ mm Hg}}\right)$$

$$\log\left(\frac{[HCO_3^-]}{0.9}\right) = 1.40$$

$[HCO_3^-] = 22.6$ mEq/liter

He has primary respiratory alkalosis due to hyperventilation, a compensatory mechanism to raise the PO_2.

19. $$7.50 = 6.10 + \log\left(\frac{31 \text{ mEq/liter}}{0.03 \times P_{CO_2}}\right)$$

$$\log\left(\frac{1033}{P_{CO_2}}\right) = 1.4$$

$P_{CO_2} = 41$ mm Hg

He has primary metabolic alkalosis without compensation. Hypokalemia (low serum K^+ level) is a common finding in primary metabolic alkalosis, particularly after diuretic therapy, which causes excessive urinary K^+ loss.

20. On a mole-for-mole basis, phosphate is a better buffer at physiologic pH than is histidine. In the serum, however, its concentration is 0.3 to 0.4 mmole per liter, which is much lower than the concentration of hemoglobin's histidine groups (48 mmoles/liter at an Hb concentration of 15 g/100 ml).

REFERENCES

Bhagavan, N. V. *Biochemistry—A Comprehensive Approach*. Philadelphia: Lippincott, 1974. Pp. 1–11, 14–18.

Davenport, H. W. *The ABC of Acid-Base Chemistry* (6th ed.). Chicago: University of Chicago Press, 1974.

Hills, A. G. *Acid-Base Balance*. Baltimore: Williams & Wilkins, 1973.

Kassirer, J. P. Serious acid-base disorders. *N. Engl. J. Med.* 291:773, 1974.

Lehninger, A. L. *Biochemistry: The Molecular Basis of Cell Structure and Function* (2nd ed.). New York: Worth, 1975. Pp. 77–80.

White, A., Handler, P., and Smith, E. L. *Principles of Biochemistry* (5th ed.). New York: McGraw-Hill, 1973. Pp. 97–106.

Winters, R. W., Engel, K., and Dell, R. B. *Acid-Base Physiology in Medicine—A Self-instruction Program* (2nd ed.). Cleveland: The London Co., 1969.

3 Structure and Properties of Polypeptides and Proteins

PEPTIDE BONDS

Peptide bonds, a type of amide bond, weld the amino end (N-terminal) of one amino acid to the carboxyl end (C-terminal) of another, as shown below. Dehydration forges this peptide linkage.

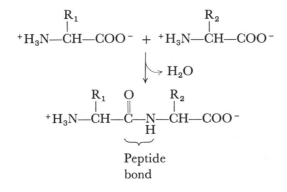

The C—N bond in the peptide linkage has partial double-bond properties that make it rigid and prevent the adjacent groups from rotating freely.

The biuret reaction gives a pinkish-purple color with peptide bonds but not with free amino acids. (Compare with the ninhydrin reaction, in which a similar color indicates the presence of free α-amino and carboxyl groups.)

Neither the C=O nor the N—H in the peptide bond can dissociate.

Problem 1

Identify the compounds below to the best of your ability and determine which have peptide bonds:

A. 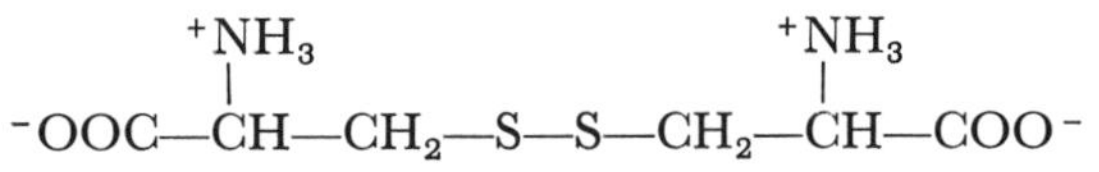 $^+NH_3$
     |
     $^-OOC—CH—CH_2—S—S—CH_2—CH—COO^-$
                                    |
                                 $^+NH_3$

B. 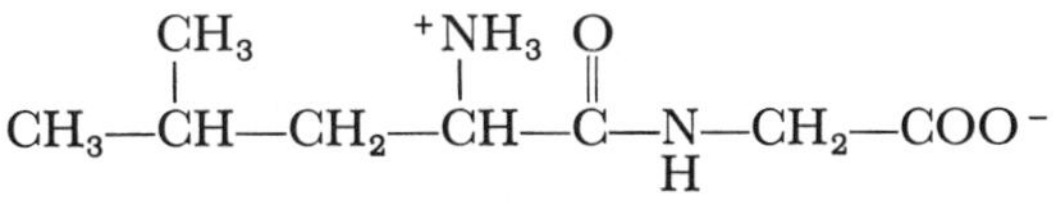 CH_3 $^+NH_3$ O
     | | ‖
     $CH_3—CH—CH_2—CH—C—N—CH_2—COO^-$
                             |
                             H

PROTEOLYSIS

Peptide-bond hydrolysis, or *proteolysis*, requires either the presence of proteolytic enzymes or heating at 110°C for 24 hours in the presence of 6 N HCl (acid hydrolysis) or NaOH (alkaline hydrolysis). Acid hydrolysis, unfortunately, destroys tryptophan and partially destroys serine, threonine, and tyrosine. Alkaline hydrolysis destroys cysteine, serine, and threonine. Since it does not damage tryptophan, alkaline hydrolysis is used in quantitative determinations of this amino acid.

AMINO ACID COMPOSITION OF POLYPEPTIDES

The distinction between the terms *oligopeptide* and *polypeptide* is somewhat arbitrary. An amino acid chain with less than 25 amino acids is called an oligopeptide, whereas a polypeptide has more than 25 amino acids. A *protein* may consist of a long polypeptide or several polypeptide subunits.

The first step in determining the amino-acid composition of a polypeptide is to measure the molecular weight of the whole, purified compound. Next, the percentage composition of each amino acid is determined, usually by automated amino acid analysis. Automated amino acid analyzers contain ion-exchange chromatography columns that separate free amino acids after complete acid hydrolysis of the polypeptide. Each amino acid is eluted from the column at a characteristic pH and is quantitated spectrophotometrically by measuring the optical density after adding ninhydrin. A second sample, which is subjected to alkaline hydrolysis, is run through the amino-acid analyzer to determine the tryptophan content. Alternatively, direct spectrophotometric tests could be used for tryptophan, which has an absorption peak in the ultraviolet range.

Finally, from the percentage composition, one calculates the number of units of each amino acid from the molecular weight, and thus the exact quantitative composition may be determined.

Problem 2

Threonine constitutes 2% of the amino-acid content of bovine insulin by weight. Calculate the molecular weight (MW) of bovine insulin, given that the MW of threonine is 119 g/mole.

STRUCTURE OF POLYPEPTIDES
Primary Structure

The amino acid sequence of a polypeptide—i.e., the order in which the amino acids occur in the chain—is defined as its *primary structure*.

Three methods are used to identify the N-terminal amino acid: (1) Sanger's method, (2) Edman's method, and (3) the leucine-aminopeptidase method.

1. In *Sanger's method*, 2,4-dinitrofluorobenzene (DNFB) binds to the N-terminal amino group of the polypeptide:

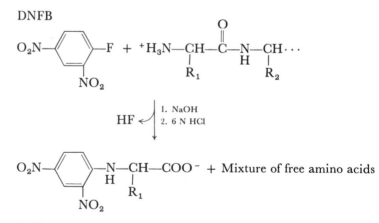

DNP-amino acid

The resultant dinitrophenyl or DNP-amino acid can be separated from the other amino acids because it is more soluble in nonpolar solvents.

2. In *Edman's method*, phenylisothiocyanate (Ph—N=C=S) combines with the N-terminal amino acid to yield a phenylthiohydantoin (PTH) compound, which may be identified by chromatography.
3. *Leucine aminopeptidase* is an exopeptidase that sequentially cleaves peptide bonds, beginning at the N-terminal end of a polypeptide. The liberated amino acids are identified one by one.

Two methods are employed for C-terminal group analyses: (1) the hydrazine method and (2) the carboxypeptidase method.

1. *Hydrazine* reacts with all amino acids whose carboxyl group is bound in peptide linkage, creating amino-acid hydrazides. Only the C-terminal amino acid is spared.
2. *Carboxypeptidases* cleave peptide bonds starting at the C-terminal end of a polypeptide. They are exopeptidases, rather than endopeptidases, because they cleave external rather than internal peptide bonds.

Endopeptidases are required in sequential analysis to break the polypeptides into oligopeptide fragments that can be separated and individually sequenced. You should memorize the sites of cleavage of two pancreatic endopeptidases: trypsin and chymotrypsin. *Trypsin* cleaves the peptide bond at the carboxyl end of the two strongly basic amino acids: arginine and lysine. *Chymotrypsin* cleaves the peptide bond at the carboxyl end of the three aromatic amino acids: phenylalanine, tyrosine, and tryptophan. Other endopeptidases, whose sites of action you need not memorize, include elastase, pepsin, and papain.

By using several endopeptidases separately on a polypeptide, different sets of oligopeptides are generated. The individual amino-acid sequences of these fragments overlap and can be fitted together like pieces of a jigsaw puzzle.

By convention, the N-terminal end of a polypeptide chain is written to the left of the C-terminal end.

Problems 3–5

Total acid hydrolysis of a pentapeptide complemented by total alkaline hydrolysis yields an equimolar mixture of five amino acids (listed alphabetically): Ala, Cys, Lys, Phe, Ser. N-terminal analysis with phenylisothiocyanate (PITC) generates PTH-serine. Trypsin digestion produces a tripeptide whose N-terminal residue is Cys and a dipeptide with Ser at its N-terminal. Chymotrypsin digestion of the above tripeptide yields Ala plus another dipeptide.

3. What is the amino acid sequence of the tripeptide?

4. What is the amino acid composition of the dipeptide derived from trypsin digestion?

5. What is the primary structure of the original pentapeptide?

Problem 6

Write the oligopeptides generated from chymotrypsin digestion of Ala-His-Tyr-Pro-Trp-Arg-Ile.

Choose the exopeptidases from the enzymes below:

A. carboxypeptidase   D. leucine aminopeptidase
B. pepsin   E. chymotrypsin
C. trypsin   F. papain

Secondary Structure

The three-dimensional geometry of the subunits of a polypeptide is termed its *secondary structure*. Common types of secondary structure include the α helix, the β or pleated sheet, the triple helix peculiar to collagen, and the random coil.

The *α helix* has 3.6 to 3.7 amino-acid subunits per each turn of its spiral. The cement that maintains this helical structure is hydrogen bonding. Groups that are able to share their hydrogen atom include $HN\diagdown$, H_2N—, ^+H_3N—, and —OH. Hydrogen-accepting groups include $\diagup C$=0, —COO^-, and —S—S—.

The α-keratins in human skin exemplify the α-helical structure. Proline and hydroxyproline disrupt the α helix because their imino groups kink the chain.

Problem 8

Which of these compounds illustrate hydrogen bonding?

A.
$$\begin{matrix} & O \\ & \| \\ & —C—O^- \cdots ^+H_3N— \end{matrix}$$

OR

B.
$$\begin{matrix} & | \\ & —C=O \cdots H_2N— \end{matrix}$$

C. $\diagup C$=0 $\cdots$ HO—

The β or *pleated sheet*, best exemplified in silk, relies on *interchain* hydrogen bonds to maintain its corrugations. In contrast, the α helix depends on *intrachain* hydrogen bonds.

The *triple helix* is peculiar to collagen. Hydrogen bonds cement the three polypeptide strands together to create tropocollagen. This tight coiling is possible because tropocollagen has a high glycine content. As the smallest amino acid, glycine allows the chains to be twisted into tight spirals. Collagen consists of many tropocollagen molecules placed end to end and parallel to each other. The staggered positioning of the ends of the tropocollagen molecules create the striations in collagen that are seen under the electron microscope.

The *random coil* lacks a repeating geometry in its subunits.

Tertiary Structure

The overall geometric conformation of a polypeptide provides its *tertiary structure*. In general, proteins may have either of two tertiary forms: fibrous or globular.

The *fibrous proteins* are elongated and water-insoluble. Their secondary structure is generally either the α-helical, pleated-sheet, or triple-helical forms, rather than random coils.

In contrast, the *globular proteins* are spherical and water-soluble, and they consist mainly of random coils with occasional stretches of α helices. The tertiary structure of globular proteins, however, is far from random, inasmuch as the primary structure of any polypeptide influences its tertiary structure. The nonpolar or hydrophobic amino acids—such as alanine, leucine, and tryptophan—tend to fold into the central area of a globular protein to exclude water as much as possible. Furthermore, these nonpolar side chains weakly attract one another. Polar and ionized side chains, on the other hand, tend to move toward the outer protein surface to form hydrogen bonds with water; they are electrostatically attracted toward side chains of opposite charge and repelled from those of like charge.

The partial double-bond nature of the peptide linkages restricts the folding of the polypeptide, and the size of the R groups also helps to govern tertiary structure. Small side chains allow tight folding of the chain, whereas bulky R groups prevent the close approach of other groups. Disulfide bonds also play a major role in determining the tertiary structure of proteins, since they rigidly link cysteine residues to one another.

Denaturation, or unfolding, of globular proteins occurs after heating or treatment with strong acids, strong bases, concentrated urea, or other agents. Proteins lose their activity after denaturation, although under certain conditions, this loss may be reversible.

Quaternary Structure	The arrangement of polypeptide chains in relation to one another in a multiple-chained protein is called the *quaternary structure*. The bonds linking these chains are all noncovalent, such as hydrogen bonds, ionic bonds, and hydrophobic bonds.

Adult hemoglobin (HbA) consists of four globin polypeptide chains—two α chains and two β chains—each bound to a separate heme group. Hemoglobin has a sigmoidal oxygen dissociation curve (see Fig. 2-2), because the binding of the initial oxygen alters the tertiary structure of the α and β chains, thereby increasing their affinity for subsequent binding to three more oxygen molecules.

PROTEIN ABNORMALITIES	Certain human disorders are associated with protein abnormalities, such as the immunoglobulinopathies and the hemoglobinopathies. The immunoglobulinopathies reflect alterations in the synthesis of γ-globulin, as seen classically in multiple myeloma.

The *hemoglobinopathies* are a class of inherited disorders that result from amino-acid substitutions in the α or β chains of the hemoglobin molecule. These disorders are detected by the resultant changes in the electrophoretic mobility of the hemoglobin.

Problem 9	How do the amino-acid substitutions below alter the charge on hemoglobin at pH 7.0? Predict whether the abnormal hemoglobin found in these hemoglobinopathies will move a greater or lesser distance than the normal form toward a positive pole during electrophoresis.

A. HbS (sickle cell) has valine instead of glutamate at position 6 on the β chain.

B. HbI has glutamate instead of lysine at position 16 on the α chain.

The altered primary structure of HbS makes deoxy-HbS insoluble and causes it to polymerize. These deoxy-HbS polymers make the erythrocyte rigid and brittle, and hemolytic anemia (so-called sickle-cell anemia) ensues.

The *thalassemias* are also hereditary hemolytic anemias, but they are characterized by the inadequate synthesis of normal α or β chains of hemoglobin. In homozygous β-thalassemia, for instance, almost no β chains are produced. To compensate for this deficiency, γ and δ chains, which are normally minor components, are overproduced. Quantitative hemoglobin electrophoresis in this case shows a lack of HbA $(\alpha_2\beta_2)$ with a great excess of HbF $(\alpha_2\gamma_2)$ and HbA$_2$ $(\alpha_2\delta_2)$.

Problem 10

The hemoglobin defect in the thalassemias lies in its:

A. primary structure
B. secondary structure
C. tertiary structure

D. quaternary structure
E. none of the above

Problem 11

The enzyme lactate dehydrogenase (LDH) is a tetramer of two subunits: M (muscle) and H (heart). How many different combinations of these two subunits can exist in an LDH tetramer?

Problem 12

In a globular protein, which amino acids would probably be positioned near the surface?

A. methionine
B. aspartate
C. phenylalanine

D. tyrosine
E. arginine
F. leucine

Problems 13–15

Choose the amino acids that interact in globular proteins as described in problems 13, 14, and 15:

A. glutamate, arginine
B. phenylalanine, tryptophan
C. aspartate, glutamate

13. Electrostatic repulsion

14. Hydrophobic attraction

15. Ionic bond

Problem 16

What is the charge on the decapeptide below at pH 7.0? You do not need a table of pK_a values nor the Henderson-Hasselbalch equation to answer this question; simply note that the pK_a of carboxyl groups is approximately 2–3, while the pK_a of amino groups is 9–10.

Ala-Met-Phe-Glu-Tyr-Val-Leu-Trp-Gly-Ile

ANSWERS

1. A. This compound, cystine, is a dimer of cysteine. Cystine is an amino acid, not a dipeptide, because a disulfide bond links the two cysteine residues.
   B. The N-terminal amino acid of this dipeptide has a branched chain, and the C-terminal is glycine.

2. $$MW = \frac{119 \text{ g/mole}}{0.02} = 5950 \text{ g/mole}$$

3. Cys-Phe-Ala
4. Ser-Lys
5. Ser-Lys-Cys-Phe-Ala
6. Ala-His-Tyr, Pro-Trp, Arg-Ile
7. A and D.
8. B and C. (Ionic bonding occurs in A.)
9. A. Compared to HbA, HbS lacks the β-carboxyl group of glutamate at position 6 of its beta chain. At pH 7.0, the glutamate group would exist as COO$^-$, i.e., with a charge of negative one. Hence, HbS is less negatively charged than HbA and will not move as far toward a positive pole.
   B. HbI lacks the ϵ-amino group of lysine at position 16. Instead, it has the β-carboxyl group of glutamate. Thus, HbI is more negatively charged than HbA, and it will move farther toward a positive pole.
10. E. Although there is a deficiency of either α or β chains, the chains that are present are normal, as are the hemoglobin tetramers that they form.
11. Five: M_4, H_1M_3, H_2M_2, H_3M_1, and H_4.
12. B, D, and E. The nonpolar residues of methionine, phenylalanine, and leucine are generally found in the central area of globular proteins.
13. C.
14. B.
15. A.
16. This decapeptide has only three dissociable groups: the α-amino group of Ala, the γ-carboxyl group of Glu, and the α-carboxyl group of Ile. At pH 7.0, both carboxyl groups will exist as COO$^-$, while the amino group will exist as $^+$NH$_3$. Hence, the net charge is -1.

REFERENCES

Barker, R. *Organic Chemistry of Biological Compounds.* Englewood Cliffs, N.J.: Prentice-Hall, 1971. Pp. 84–137.

Bhagavan, N. W. *Biochemistry—A Comprehensive Review.* Philadelphia: Lippincott, 1974. Pp. 32–57.

Lehninger, A. L. *Biochemistry: The Molecular Basis of Cell Structure and Function* (2nd ed.). New York: Worth, 1975. Pp. 95–117, 125–154.

Light, A. *Protein Structure and Function.* Englewood Cliffs, N.J.: Prentice-Hall, 1974.

Wetlaufer, D. B., and Ristow, S. Acquisition of three-dimensional structure of proteins. *Annu. Rev. Biochem.* 42:135, 1973.

White, A., Handler, P., and Smith, E. L. *Principles of Biochemistry* (5th ed.). New York: McGraw-Hill, 1973. Pp. 108–163.

4 Enzymes

Enzymes can best be defined as catalysts with a high degree of specificity for a certain substrate or class of substrates. As a catalyst, they lower the activation energy of chemical reactions (see Ch. 6). Like other catalysts, enzymes in very low concentrations enhance both the forward and reverse reaction rates without themselves being consumed. Unlike the metallic and inorganic catalysts, however, each enzyme can act on only one substrate or on a family of structurally similar substrates.

All enzymes contain a polypeptide or protein. Some also have nonprotein prosthetic groups, such as heme, heavy metals, or coenzymes. Coenzymes are nonprotein, organic molecules that assist enzymes in transferring certain groups (see Ch. 5).

ENZYME NOMENCLATURE

In general, the name of an enzyme consists of two parts: first, the name of the substrate, or occasionally that of the product, is stated; the second portion of the name describes the type of reaction.

The International Union of Biochemists (IUB) has assigned a recommended name for each enzyme to replace its historical name or names. Some texts have adopted these recommended names, whereas others continue to use the former names. This text will generally use the recommended names. The IUB has also assigned a systematic name for each enzyme, which is often too long and cumbersome to be adopted for general usage.

ENZYME TYPES

Aldolase: Cleaves a carbon-carbon bond to create an aldehyde group.

Carboxylase: Adds CO_2 or HCO_3^- to its substrate to form a carboxyl group.

Decarboxylase: Cleaves a carboxyl group, e.g., from α-keto acids, liberating it as CO_2.

Dehydrogenase: Removes hydrogen atoms from its substrate.

Esterase: Hydrolyzes ester linkages to form an acid and an alcohol.

Hydratase: Adds water to a carbon-carbon double bond without breaking the bond or, conversely, removes water to create a double bond.

Hydrolase: Adds water to break a bond (hydrolysis). The suffix "ase" alone often denotes a hydrolase; e.g., sucrase hydrolyzes sucrose.

Hydroxylase: Incorporates an oxygen atom from O_2 into its substrate to create a hydroxyl group.

Isomerase: Converts between *cis* and *trans* isomers, D and L isomers, or aldose and ketose.

Kinase: Transfers a phosphate group from a high-energy phosphate compound, such as ATP, to its substrate (in contrast, a phosphorylase adds inorganic phosphate, P_i, to its substrate).

Ligase: Joins two molecules together using the energy released from hydrolyzing a pyrophosphate bond of a high-energy phosphate compound; also called synthetase.

Lyase: Adds groups to double bonds or removes groups to create double bonds, other than by hydrolysis.

Mutase: Shifts the position of a group, e.g., a methyl group, within a single molecule.

Oxidase: Adds O_2 to hydrogen atoms removed from the substrate (which is thereby oxidized) to generate H_2O, H_2O_2, or O_2^- (superoxide).

Oxygenase: Incorporates molecular O_2 into its substrates.

Peptidase: Hydrolyzes peptide bonds to yield free amino acids and peptides.

Phosphatase: Hydrolyzes substrates, such as phosphoric esters, to liberate inorganic phosphate (P_i; at physiologic pH, a mixture of HPO_4^{-2} and $H_2PO_4^-$).

Phosphorylase: Adds inorganic phosphate (P_i) to split a bond (phosphorolysis).

Reductase: Catalyzes the reduction of its substrate, i.e., adds hydrogen atoms.

Sulfatase: Hydrolyzes substrates, such as sulfuric-acid esters, to liberate sulfate.

Synthase: Joins two molecules together without hydrolyzing a pyrophosphate bond (in contrast, ligase or synthetase requires the hydrolysis of such a bond).

Synthetase: Same as ligase.

Transaminase: Transfers amino groups from an amino acid to a keto acid (also known as aminotransferase).

Transferase: Transfers groups other than hydrogen atoms, such as phosphate groups for phosphotransferases or methyl groups for methyltransferases, from one molecule to another.

Problems 1–6

For each reaction given below, name the enzyme that catalyzes it:

1.

$$CH_3-\overset{\overset{O}{\|}}{C}-COO^- + CO_2 + ATP \longrightarrow {}^-OOC-CH_2-\overset{\overset{O}{\|}}{C}-COO^- + ADP + P_i$$

Pyruvate Oxaloacetate

2.

Methylmalonyl-CoA   Succinyl-CoA

3.

$$-\overset{\overset{H}{|}}{C}=\overset{\overset{H}{|}}{C}- \longrightarrow -\overset{\overset{H}{|}}{C}=\overset{\overset{H}{|}}{C}-$$

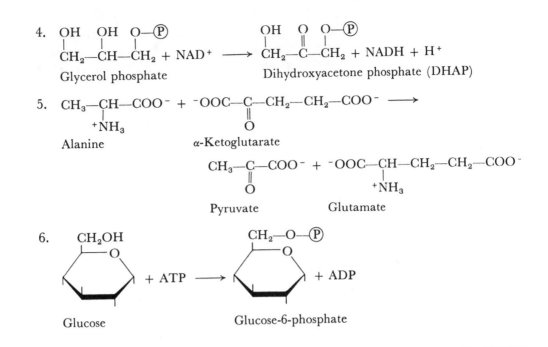

4. Glycerol phosphate + NAD$^+$ $\longrightarrow$ Dihydroxyacetone phosphate (DHAP) + NADH + H$^+$

5. CH$_3$—CH—COO$^-$ + $^-$OOC—C—CH$_2$—CH$_2$—COO$^-$ $\longrightarrow$
        | $^+$NH$_3$ ‖ O

   Alanine   α-Ketoglutarate

   CH$_3$—C—COO$^-$ + $^-$OOC—CH—CH$_2$—CH$_2$—COO$^-$
        ‖ O | $^+$NH$_3$

   Pyruvate   Glutamate

6. Glucose + ATP $\longrightarrow$ Glucose-6-phosphate + ADP

Problems 7–9

Draw the structures of the products for the reactions below and describe the reaction:

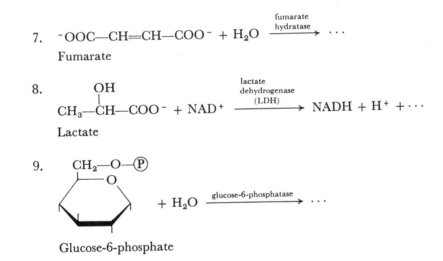

7. $^-$OOC—CH=CH—COO$^-$ + H$_2$O $\xrightarrow{\text{fumarate hydratase}}$ $\cdots$

   Fumarate

8. CH$_3$—CH—COO$^-$ + NAD$^+$ $\xrightarrow{\text{lactate dehydrogenase (LDH)}}$ NADH + H$^+$ + $\cdots$
         | OH

   Lactate

9. Glucose-6-phosphate + H$_2$O $\xrightarrow{\text{glucose-6-phosphatase}}$ $\cdots$

ENZYME KINETICS

An enzyme (E) catalyzes a reaction by combining with its substrate (S) to create an enzyme-substrate complex (ES). The specificity of each enzyme for one or for several substrates compares to the fit of a key (the substrate) into a lock (the active or catalytic site of the enzyme).

    Factors that alter enzyme-substrate binding include pH and temperature. Because of the charge that may exist on ionizable R groups near the active site at high or low pH values, ionic bonds or electrostatic attractions or repulsions may be present that can enhance, diminish, or prevent substrate binding. Hence, each enzyme has an *optimal pH value* at which it has maximal activity. This optimal pH is often near the pH of the tissue that contains the enzyme; for example, pepsin secreted into the stomach has an optimal pH of about 2.0

(gastric pH is 2.0 to 3.0); pancreatic α-amylase, when secreted into the duodenum, has an optimal pH of about 7.0 (intestinal pH is 6.0 to 7.0); and alkaline phosphatase in bone has an optimal pH of 9 to 10 (bone pH is above 7.4).

Raising the temperature increases the reaction rate (provided the higher temperature does not denature the enzyme), because it increases the kinetic energy of the molecules, thus allowing more frequent collisions between the enzyme and the substrate.

Reaction Order

For the reaction below, let K be the *rate constant*, R the *reaction order*, and V the *rate*:

$$S \xrightarrow{K} P$$
$$V = K[S]^R \tag{4-1}$$

For a *zero-order reaction* with respect to $[S]$, the rate is constant regardless of $[S]$:

$$V = K[S]^0 = K \qquad \text{(zero order)} \tag{4-2}$$

In *first-order reactions* with respect to $[S]$, the rate is proportional to $[S]$, while in *second-order reactions*, the rate is proportional to $[S]^2$:

$$V = K[S] \qquad \text{(first order)} \tag{4-3}$$
$$V = K[S]^2 \qquad \text{(second order)} \tag{4-4}$$

Problem 10

Estimate the reaction order with respect to $[S]$ for tracings A–D in Figure 4-1.

Michaelis-Menten Equation

The enzyme-substrate complex (ES) has two fates: S can be converted to P (product) or ES can dissociate back to $E + S$, as represented below:

$$E + S \underset{K_{-1}}{\overset{K_1}{\rightleftharpoons}} ES \xrightarrow{K_2} P \tag{4-5}$$

Figure 4-1

From this graph of initial velocity (V) vs initial substrate concentration ($[S]_0$), estimate the reaction order with respect to $[S]$ for curves A–D (Problem 10).

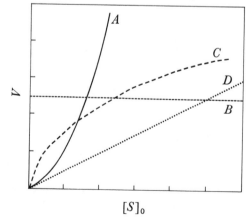

$[S]_0$

Michaelis and Menten reasoned that by measuring the initial reaction velocity (V) rather than later rates (when much of S had been converted to P), they could assume that $[S]$ remained at $[S]_0$, the initial substrate concentration, and that $[P]$ was zero. They then set up mathematical expressions for the rate of all three reactions in Equation 4-5, defining V_{max} as the velocity extrapolated to infinitely high $[S]$ and K_m as $(K_{-1} + K_2)/K_1$. If $[ES]$ remains constant (the steady-state assumption), then these rate expressions become:

$$V = \frac{V_{max}[S]_0}{K_m + [S]_0} \qquad\qquad (4\text{-}6)$$

Equation 4-6 states the *Michaelis-Menten equation*, where K_m, the *Michaelis constant*, represents the substrate concentration at which the reaction rate is one-half V_{max}. It is *not* an equilibrium constant. Furthermore, since K_1, K_2, and K_{-1} cannot be empirically measured, K_m cannot be calculated from them; instead, K_m must be determined experimentally after measuring V_{max} and V at various substrate concentrations.

With certain enzymes, the dissociation of ES to an enzyme-product complex (EP) is the rate-limiting step. In such circumstances, K_m becomes the dissociation content for the enzyme-substrate complex (ES).

Enzymes with several substrates will possess a different K_m for each substrate. Both K_m and V_{max} for each enzyme will vary with changes in pH and temperature.

Problem 11

Use the Michaelis-Menten equation to complete the table below. Graph your results in Figure 4-2, and determine the order of the reaction (note that since K_m and V_{max} are constants, your curve will have the same slope as one of V versus $[S]_0$).

$\dfrac{[S]_0}{K_m}$	$\dfrac{V}{V_{max}} \times 100$
$\frac{1}{2}$	33
1	. . .
2	. . .
3	. . .
10	. . .

Problem 12

The K_m of hexokinase for glucose is 0.15 mM, whereas its K_m for fructose is 1.5 mM. Assume V_{max} is the same for both substrates.

A. Write the reactions of glucose and fructose that are catalyzed by hexokinase.

B. Calculate V as a percentage of V_{max} for each substrate when $[S]_0 = 0.15$ mM, 1.5 mM, and 15 mM.

C. Which substrate does hexokinase have a greater affinity for?

Figure 4-2

Graph your results from Problem 11 here to determine the reaction order.

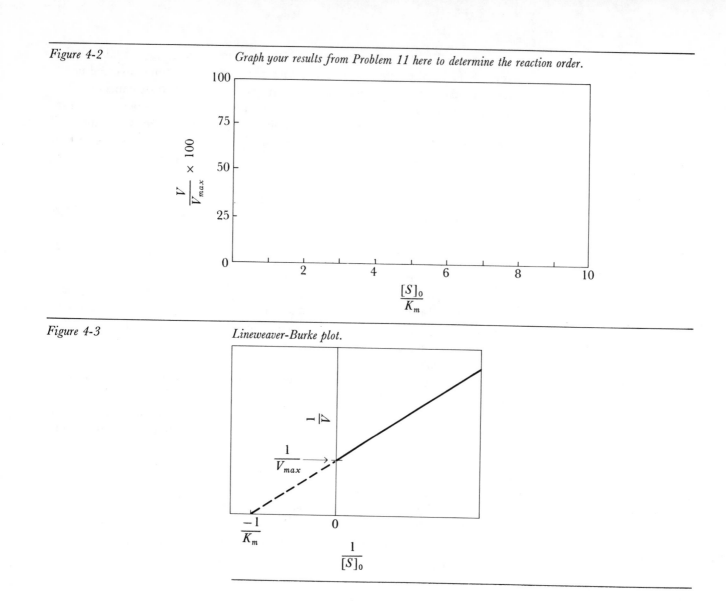

Figure 4-3   *Lineweaver-Burke plot.*

By taking the reciprocal of the Michaelis-Menten equation, one gets the *Lineweaver-Burke equation*:

$$\frac{1}{V} = \frac{K_m}{V_{max}[S]_0} + \frac{1}{V_{max}} \tag{4-7}$$

The double-reciprocal plot of $1/V$ versus $1/[S]_0$ is quite useful, because its *y*-intercept is $1/V_{max}$ while its *x*-intercept is $-1/K_m$, as illustrated in Figure 4-3.

ENZYME INHIBITORS

There are two classes of enzyme inhibitors: reversible and irreversible. *Irreversible inhibitors* bind covalently to enzymes and dissociate very slowly, as indicated by the thickness of the arrows below:

$$E + I \rightleftharpoons EI$$

38   4. Enzymes

Figure 4-4

Effect of competitive inhibitor concentration, [I], on K_m and V_{max} of a hypothetical enzyme (see Problem 13).

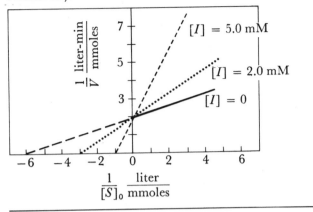

Because the irreversible inhibitors permanently inactivate enzymes, the Michaelis-Menten equation, which requires a constant enzyme concentration, does not apply.

Toxic organophosphorous compounds, such as certain insecticides and diisopropyl fluorophosphate (DFP), act as irreversible inhibitors by clinging to the active site of human acetylcholinesterase, which leads to a toxic accumulation of acetylcholine.

Cyanide and sulfide bind to the iron atom of cytochrome oxidase, causing irreversible inhibition. Not all irreversible inhibitors, however, bind to the active site of enzymes.

Reversible inhibitors bind noncovalently to enzymes through hydrogen bonds or ionic bonds. Unlike the case of irreversible inhibitors, the reactions affected obey Michaelis-Menten kinetics.

Reversible enzyme inhibitors may have either of two modes of action: competitive inhibition or noncompetitive inhibition.

Competitive inhibitors are chemically analogous to the substrate and bind to the active sites of enzymes so that they compete with the substrate for enzyme binding. Competitive inhibition is reversible, since high substrate concentrations will overcome the effect of the inhibitor. In such reactions, the V_{max} will be reached, but K_m increases with increasing inhibitor concentration because the inhibitor reduces substrate binding to the catalytic site. Thus, the degree of inhibition depends upon the ratio $[I]/[S]$, where $[I]$ represents the inhibitor concentration. Figure 4-4 shows the characteristic increase in K_m with no change in V_{max} in the presence of a competitive inhibitor.

Calculate K_m and V_{max} for the hypothetical enzyme with the competitive inhibitor in Figure 4-4 when $[I] = 0$, 2.0, and 5.0 mM.

Noncompetitive inhibitors bind to enzymes in areas other than the active site. Unlike competitive inhibitors, they do not resemble the substrate. The degree of inhibition depends upon $[I]$; raising $[S]$ will not overcome the inhibition. Hence, in contrast to competitive inhibition, V_{max} decreases. As is

Figure 4-5

Effect of noncompetitive inhibitor concentration, $[I]$, on K_m and V_{max} of a hypothetical enzyme (see Problem 14).

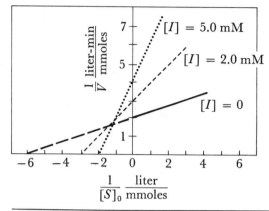

the case with competitive inhibition, however, K_m usually increases (Fig. 4-5), but it occasionally remains the same.

Heavy metals such as Hg^{+2} and Pb^{+2} bind to the sulfhydryl groups in enzymes and inhibit noncompetitively.

Irreversible inhibitors are not classified as competitive or noncompetitive.

Problem 14

Calculate K_m and V_{max} for the hypothetical enzyme with the noncompetitive inhibitor shown in Figure 4-5 when $[I] = 0$, 2.0, and 5.0 mM.

Problems 15–20

Classify each inhibitor of the following reactions as reversible or irreversible. If reversible, decide whether it is competitive or noncompetitive. Assign names to the enzymes that catalyze these reactions where they are not given.

15. $^-OOC—CH_2—CH_2—COO^- + FAD^+ \rightleftharpoons$
Succinate
$$^-OCC—CH=CH—COO^- + FADH_2$$
Fumarate

Inhibitor is malonate, $^-OOC—CH_2—COO^-$.

16. Cytochrome oxidase contains iron which reacts as follows:

$$2Fe^{+2} + 2H^+ + \tfrac{1}{2}O_2 \longrightarrow 2Fe^{+3} + H_2O$$

Inhibitor is cyanide, $C\equiv N^-$.

17. OH
$+ O_2 + H_2O \rightleftharpoons$
OH

Hypoxanthine
(enol form)

Xanthine
(enol form)

Inhibitor is allopurinol,

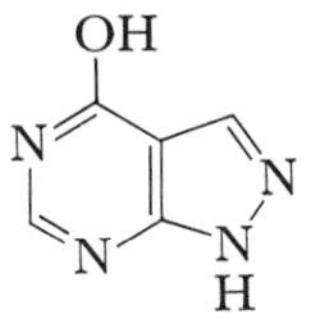

18. Lead poisoning causes hypochromic microcytic anemia because it inhibits several enzymes involved in synthesizing hemoglobin. What type of inhibition is this? (Do not try to name the enzymes.)

19.

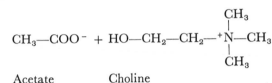

Acetylcholine

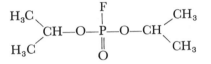

Acetate   Choline

Inhibitor is DFP,

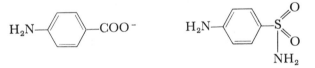

20. Certain bacteria cannot utilize pre-formed folic acid. Instead, they must synthesize their own folic acid by joining *p*-amino benzoate (PABA) to a pteridine compound and then adding glutamate to yield pteroylglutamic acid (folic acid). Sulfanilamide inhibits enzymes in this synthesis (do not try to name the enzymes). What kind of inhibitor is sulfanilamide?

PABA   Sulfanilamide

ALLOSTERIC ENZYMES

The word *allosteric* means "another site." By definition, an *allosteric enzyme* has a regulatory site, which differs from its catalytic site, that binds allosteric effectors, also called modulators or modifiers. Virtually all allosteric enzymes have multiple polypeptide subunits. Usually, noncovalent bonds are formed between the effector and the enzyme.

Positive or *stimulatory effectors* enhance substrate binding, whereas *negative effectors* reduce substrate binding. Upon binding to the regulatory site, effectors

Figure 4-6

Rectangular-hyperbolic plot that is characteristic of the reactions of nonallosteric enzymes (compare Problem 11).

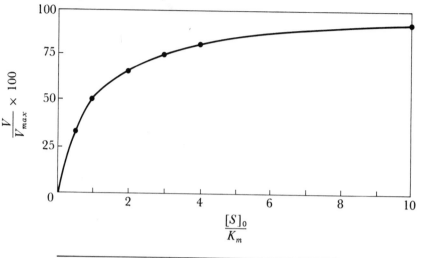

change the quaternary structure of the allosteric enzyme, thereby altering the substrate-binding properties of the catalytic site.

Unlike a competitive inhibitor, an allosteric modifier need not resemble its enzyme's substrate, although it may do so occasionally.

Allosteric enzymes are often strategically placed at the first, or *committed step*, of a long metabolic pathway. The final endproduct of the pathway then acts as a negative modulator for that enzyme; this is termed *endproduct* or *feedback inhibition*. Such is the case for aspartate transcarbamoylase (ATCase), a transferase that controls the rate of pyrimidine synthesis in humans. ATCase adds carbamoyl phosphate to aspartate to yield *N*-carbamoylaspartate, an intermediate in pyrimidine biosynthesis. Cytidine triphosphate (CTP), an endproduct of pyrimidine synthesis, acts as a negative allosteric modulator. When the CTP level becomes high, the enzyme is inhibited, thus shutting off further synthesis. On the other hand, the purine nucleoside phosphates (such as ATP) allosterically stimulate ATCase, primarily by preventing the binding of CTP.

Regardless of the presence or absence of its effectors, an allosteric enzyme does not obey Michaelis-Menten kinetics. When V versus $[S]_0$ is graphed for an allosteric enzyme reaction, the curve deviates from the rectangular hyperbola of a nonallosteric enzymatic reaction, shown in Figure 4-6. Most allosteric enzyme reactions display a sigmoidal curve, which indicates that the binding of one substrate facilitates the binding of additional substrate molecules to other active sites. This sigmoidal curve also appears for the oxygen-binding reaction of the nonenzymatic protein, hemoglobin (see Fig. 2-2).

PHOSPHORYLATED REGULATORY ENZYMES

Allosteric enzymes depend upon nonconvalent binding of their modifiers, whereas a second type of regulatory enzyme, the *phosphorylated regulatory enzymes*, depends upon *covalent* binding between phosphate and the enzyme to modify the enzyme's activity. For example, active (dephospho) glycogen synthase is phosphorylated by protein kinase to the inactive (phospho) form.

On the other hand, the reverse process occurs with glycogen phosphorylase: inactive (dephospho) glycogen phosphorylase is transformed into the active (phospho) form by phosphorylation.

ISOZYMES

Enzymes that contain multiple polypeptide subunits of two or more types and which exist in several different forms are termed *isozymes*. Lactate dehydrogenase (LDH), for example, is a tetramer of H (heart) and M (muscle) subunits. Five isozymes of LDH exist: H_4, H_3M, H_2M_2, HM_3, and M_4. These isozymes differ from one another with respect to the K_m and V_{max} values for the reaction with lactate.

ENZYME ACTIVITY

The *international unit (IU) of enzyme activity* is defined as the quantity of enzyme needed to transform 1.0 micromole of substrate to product per minute at 30°C and optimal pH. To determine the enzyme activity in a blood sample, for instance, one must measure the rate of the enzymatically catalyzed reaction with excess substrate present to insure that the enzyme has been saturated and that the rate depends only on the enzyme's activity.

A convenient method for measuring the reaction rate is to couple the reaction to the conversion of NADH to NAD^+ or of NADPH to $NADP^+$. NADH and NADPH absorb light at 340 nm to a much greater extent than their oxidized counterparts. Thus, their removal or production can be followed by measuring the optical density (OD) spectrophotometrically.

Problem 21

You couple an assay of SGPT (serum glutamate-pyruvate transaminase) to the removal of NADH. To 0.20 ml of serum, you add your reagents, all present in excess. The OD drops 0.63 unit in the two minutes required to complete the reaction. Given that one IU of enzyme activity is equivalent to 2.1 OD units, what is the SGPT activity in IU/ml?

ANSWERS

1. Carbon dioxide joins pyruvate to form oxaloacetate. Hence, the enzyme is called pyruvate carboxylase.
2. Methylmalonyl-CoA mutase. It is a mutase because it shifts the position of the (C=O)CoA group within the same molecule. (CoA = coenzyme A.)
3. *Cis-trans* isomerase.
4. Glycerolphosphate dehydrogenase (P represents phosphate).
5. Transaminase or aminotransferase. The amino group of alanine is transferred to α-ketoglutarate to form glutamate. Hence, this is a transamination reaction. Alanine-pyruvate transaminase is one of its several names.
6. Kinase. Two enzymes may catalyze this reaction: hexokinase, which can use various sugars as its substrate, or glucokinase, which is specific for glucose.

7.
$$^-OOC—\overset{\overset{\displaystyle OH}{|}}{C}H—CH_2—COO^-$$

Malate. A hydratase adds H_2O across a double bond.

8.

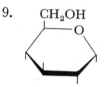

Pyruvate. Lactate dehydrogenase removes the hydrogen atoms from lactate's hydroxyl group and α carbon to generate a keto group.

9.

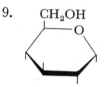

Glucose plus P_i. A phosphatase removes this phosphate group.

10. A. Second-order.
    B. Zero-order.
    C. Mixed zero- and first-order.
    D. First-order.

11.

$\frac{[S]_0}{K_m}$	$\frac{V}{V_{max}} \times 100$
$\frac{1}{2}$	33
1	50
2	67
3	75
10	91

The reaction order is mixed zero- and first-order; compare this with the graph in Figure 4-6, a rectangular hyperbola, which is characteristic of nonallosteric enzymes.

12. A. Glucose + ATP $\rightleftharpoons$ Glucose-6-P + ADP
    Fructose + ATP $\rightleftharpoons$ Fructose-6-P + ADP
    B. $[S]_0 = 0.15$ mM:

For glucose, V is 50% V_{max} because $[S]_0 = K_m$
For fructose,

$$V = \frac{V_{max}(0.15)}{1.5 + (0.15)} = 9.1\% \ V_{max}$$

$[S]_0 = 1.5$ mM:
For glucose,

$$V = \frac{V_{max}(1.5)}{0.15 + (1.5)} = 91\% \ V_{max}$$

For fructose, V is 50% V_{max} because $[S]_0 = K_m$

$[S]_0 = 15$ mM:
For glucose,

$$V = \frac{V_{max}(15)}{0.15 + (15)} = 99\% \ V_{max}$$

For fructose,

$$V = \frac{V_{max}(15)}{1.5 + (15)} = 91\% \; V_{max}$$

    C. Glucose. At low values of $[S]_0$, such as 0.15 mM, hexokinase phosphorylates glucose much more rapidly than it does fructose. At high values of $[S]_0$, such as 15 mM, either substrate saturates the active sites of hexokinase and both reactions proceed at rates near V_{max}.

13. For all inhibitor concentrations, $V_{max} = 0.50$ mmoles/liter-min
    When $[I] = 0$, $K_m = 0.167$ mM
    When $[I] = 2.0$ mM, $K_m = 0.333$ mM
    When $[I] = 5.0$ mM, $K_m = 1.0$ mM

14. When $[I] = 0$, $V_{max} = 0.50$ mmoles/liter-min and $K_m = 0.167$ mM
    When $[I] = 2.0$ mM, $V_{max} = 0.33$ mmoles/liter-min and $K_m = 0.333$ mM
    When $[I] = 5.0$ mM, $V_{max} = 0.25$ mmoles/liter-min and $K_m = 0.500$ mM

15. Succinate dehydrogenase. Because malonate resembles succinate, you should predict that it reversibly and competitively inhibits.

16. Irreversible inhibition.

17. Allopurinol, a drug used to treat gout, resembles hypoxanthine sufficiently closely to inhibit xanthine oxidase reversibly and competitively. (You are justified to have named this enzyme hypoxanthine oxidase; it is called xanthine oxidase because it also oxidizes xanthine to uric acid. Another fact that you could not have predicted is that allopurinol can be oxidized to oxypurinol, which inhibits this enzyme noncompetitively.)

18. Reversible, noncompetitive inhibition.

19. Irreversible inhibition of acetylcholinesterase. This is an esterase, because it hydrolyzes an ester. One might also name it acetycholine hydrolase, since the bond is hydrolyzed.

20. Reversible, competitive inhibitor.

21. SGPT activity:

$$\frac{IU}{ml} = \frac{(0.63 \; \text{OD unit})}{(2.0 \; \text{min} \times 0.20 \; \text{ml})} \times \frac{(1.0 \; \text{IU})}{(2.1 \; \text{OD units})} = 0.75 \; \text{IU/ml}$$

REFERENCES

Christensen, H. N., and Palmer, G. A. *Enzyme Kinetics—A Learning Program for Students of the Biological and Medical Sciences* (2nd ed.). Philadelphia: Saunders, 1974.

Lehninger, A. L. *Biochemistry: The Molecular Basis of Cell Structure and Function* (2nd ed.). New York: Worth, 1975. Pp. 183–201, 207–208, 234–246.

Sigman, D. S., and Mooser, G. Chemical studies of enzyme active sites. *Annu. Rev. Biochem.* 44:889, 1975.

White, A., Handler, P., and Smith, E. L. *Principles of Biochemistry* (5th ed.). New York: McGraw-Hill, 1973. Pp. 213–275.

5 Vitamins and Coenzymes

A *coenzyme* is a nonprotein organic molecule that binds to an enzyme to aid in the transfer of specific functional groups. Usually, it binds loosely and can be easily separated from its enzyme, but when it binds tightly, it is considered to be a *prosthetic group* of the enzyme.

A *cofactor* differs from a coenzyme only in that it is a metallic ion rather than an organic molecule. Examples include Fe^{+2} in the cytochromes, Mg^{+2} for enzymes utilizing ATP, Zn^{+2} in lactate dehydrogenase, Mo^{+6} in xanthine oxidase, and Cu^{+2} in cytochrome oxidase.

In the process of transfering functional groups, both coenzymes and cofactors often change their own structure or valence and must be later returned to their original form.

For humans, a *vitamin* is an organic molecule required for certain metabolic functions that must be supplied in very small amounts (less than 50 mg/day) because we either cannot synthesize it or cannot synthesize enough to meet our needs. This definition excludes inorganic compounds, such as metals and minerals, and essential nutrients required in large amounts, such as amino acids, glucose, and triglycerides.

The *minimal daily requirement* (MDR) of a vitamin is the minimal oral intake necessary to prevent the symptoms and signs of vitamin deficiency from appearing. Meeting but not exceeding the MDR may still allow the biochemical abnormalities of vitamin deficiency to be present. Therefore, the MDR is not in itself an adequate vitamin intake.

The *recommended daily allowance* (RDA) of a vitamin represents an estimate of the adequate vitamin intake for the majority of healthy Americans in each age bracket. Meeting the RDA will guarantee an adequate vitamin supply for all but a small minority of Americans. An intake less than the RDA, however, does not necessarily cause a vitamin shortage.

WATER-SOLUBLE VITAMINS AND COENZYMES

The *water-soluble vitamins*, which include all B vitamins and vitamin C, act as coenzymes or coenzyme precursors.

Glossitis, the loss of tongue papillae, occurs in five of the B-vitamin deficiency states (folic acid, niacin, riboflavin, pyridoxine, and cyanocobalamin) as well as in iron deficiency. Friedman and Hodges (1977) showed that 14% of 117 hospitalized patients had glossitis. Hence, these B-vitamin deficiency states are seen every day in hospitals in the U.S.A.

In the following structural formulas, an asterisk (*) designates the reactive site, insofar as it has been determined.

Thiamine Pyrophosphate (TPP)

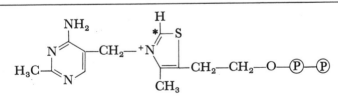

Thiamine pyrophosphate (TPP)

TPP is an ester of thiamine or vitamin B_1. It should not be confused with thymine, a pyrimidine.

The essential role of TPP is to serve as a coenzyme in the decarboxylation of α-keto acids and keto sugars. TPP is one of the coenzymes required in the pyruvate dehydrogenase reaction that links carbohydrate metabolism to the tricarboxylic-acid cycle (TCA cycle). The overall reaction involves the hydrogenation and then the dehydrogenation of the α-keto group of pyruvate along with removal of its carboxyl group:

$$^-OOC\!-\!\overset{\overset{\displaystyle O}{\|}}{C}\!-\!CH_3 + NAD^+ + CoA\!-\!SH \longrightarrow$$

$$CoA\!-\!S\!-\!\overset{\overset{\displaystyle O}{\|}}{C}\!-\!CH_3 + CO_2 + NADH$$

This reaction occurs in five steps:

1. $\quad TPP + ^-\mathbf{OOC}\!-\!\overset{\overset{\displaystyle O}{\|}}{C}\!-\!CH_3 \longrightarrow TPP\!-\!\overset{\overset{\displaystyle OH}{|}}{CH}\!-\!CH_3 + \mathbf{CO_2}$

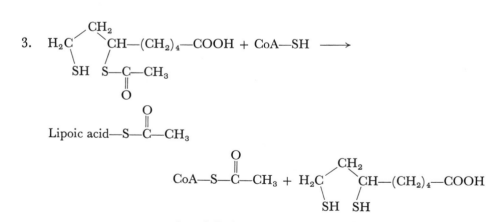

2.

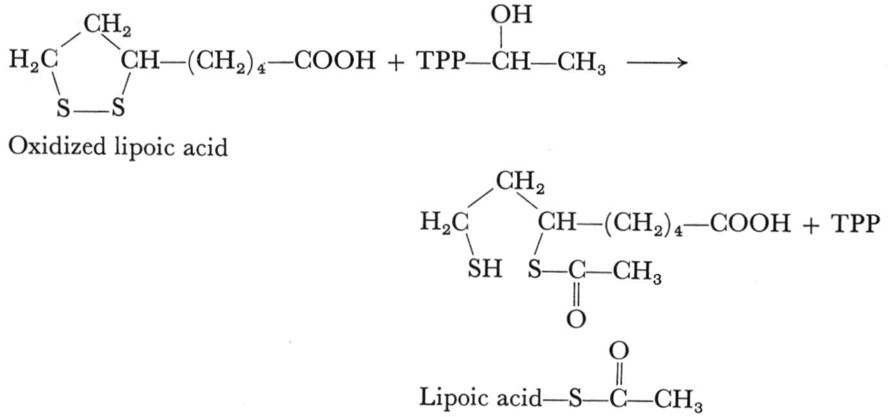

Oxidized lipoic acid

Lipoic acid—S—C—CH₃

3.

Lipoic acid—S—C—CH₃

Acetyl-CoA   Reduced lipoic acid

4.

$$\text{H}_2\text{C} \underset{\underset{\text{SH}}{|}}{\overset{\overset{\displaystyle \text{CH}_2}{\diagup}}{}} \text{CH} \underset{\underset{\text{SH}}{|}}{} -(\text{CH}_2)_4-\text{COOH} + \text{FAD} \longrightarrow$$

Reduced lipoic acid

$$\text{H}_2\text{C} \overset{\overset{\displaystyle \text{CH}_2}{\diagup}}{} \text{CH} -(\text{CH}_2)_4-\text{COOH} + \text{FADH}_2$$
$$\text{S}-\text{S}$$

Oxidized lipoic acid

5. $\text{FADH}_2 + \text{NAD}^+ \longrightarrow \text{FAD} + \text{NADH} + \text{H}^+$

TPP performs the initial decarboxylation of pyruvate. Oxidized lipoic acid then removes the —CHOH—CH$_3$ group from TPP and oxidizes it. Coenzyme A (CoA—SH) in turn removes this group from reduced lipoic acid, generating acetyl-CoA, which enters the TCA cycle. FAD oxidizes the reduced lipoic acid to regenerate oxidized lipoic acid, and finally, NAD$^+$ oxidizes FADH$_2$ to yield NADH and H$^+$.

TPP is also used in the transketolase reaction of the hexose-monophosphate shunt to transfer a —(C=O)—CH$_2$OH group between two sugar phosphates.

Signs of thiamine deficiency appear in people whose caloric intake is disproportionately high compared to their thiamine intake. Such an imbalance occurs endemically in certain areas of Asia where people subsist largely on polished, milled rice (the processing of which removes the thiamine) and also in chronic alcoholics who eat little food (alcoholic beverages provide calories but not thiamine). Thus, the RDA for thiamine is stated in proportion to the caloric intake.

A moderate thiamine deficiency impairs carbohydrate metabolism in neurons, producing peripheral neuropathy ("dry" beriberi). Severe thiamine deficiency impairs carbohydrate metabolism in the heart and blood vessels, causing high-output congestive heart failure ("wet" beriberi).

Thiamine deficiency can cause sudden dementia, ataxia, and ophthalmoplegia (Wernicke's encephalopathy), which may become irreversible unless the condition is quickly treated with thiamine.

The principal dietary sources of thiamine are meats, beans, peas, and grains.

Flavin Mononucleotide (FMN) and Flavin-Adenine Dinucleotide (FAD)

Riboflavin, or vitamin B$_2$, is phosphorylated in the intestine to generate FMN. ATP then adds AMP to FMN to yield FAD:

FMN + ATP $\longrightarrow$ FAD + PP$_i$

where PP$_i$ indicates inorganic pyrophosphate.

Both FAD and FMN transfer hydrogen atoms and electrons, utilizing the two nitrogen atoms designated by asterisks in the formula on page 50. They are used by flavin-linked enzymes as in the oxidative deamination of amino acids, the β-oxidation of fatty acids, purine catabolism, and oxidative phosphorylation.

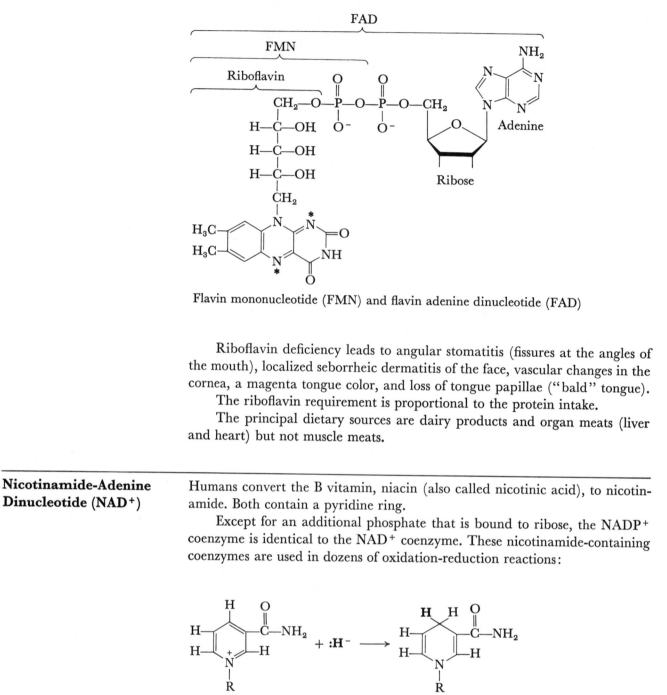

Flavin mononucleotide (FMN) and flavin adenine dinucleotide (FAD)

Riboflavin deficiency leads to angular stomatitis (fissures at the angles of the mouth), localized seborrheic dermatitis of the face, vascular changes in the cornea, a magenta tongue color, and loss of tongue papillae ("bald" tongue).

The riboflavin requirement is proportional to the protein intake.

The principal dietary sources are dairy products and organ meats (liver and heart) but not muscle meats.

Nicotinamide-Adenine Dinucleotide (NAD⁺)

Humans convert the B vitamin, niacin (also called nicotinic acid), to nicotinamide. Both contain a pyridine ring.

Except for an additional phosphate that is bound to ribose, the $NADP^+$ coenzyme is identical to the NAD^+ coenzyme. These nicotinamide-containing coenzymes are used in dozens of oxidation-reduction reactions:

Nicotinamide group of NAD⁺   Hydride ion   Nicotinamide group of NADH

Humans convert a fraction of their dietary tryptophan to nicotinamide. Thus, the combined niacin-tryptophan intake determines whether enough nicotinamide can be supplied through the diet.

Niacin-deficiency disease, or pellagra, develops in people whose niacin-tryptophan intake is low compared to their caloric intake. Pellagra is classically a disease of people who subsist mainly on corn, which is low in both niacin and tryptophan. Signs of pellagra include dermatitis, diarrhea, and dementia (the three Ds), and loss of tongue papillae.

Major food sources of niacin are meats and nuts.

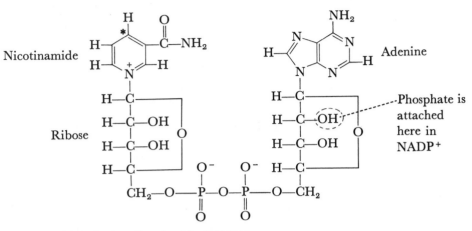

Nicotinamide-adenine dinucleotide (NAD⁺)

Coenzyme A (CoA)

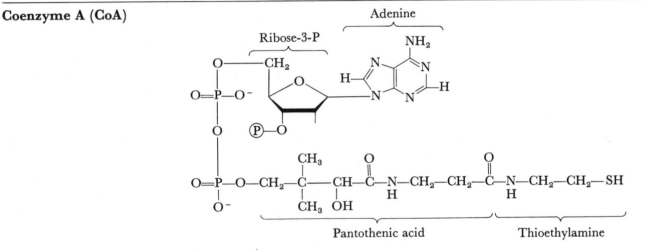

Coenzyme A

Pantothenic acid, a B vitamin, combines with ATP and cysteine in the liver to generate CoA—SH.

CoA—SH transfers acyl groups, R—(C=O)—, such as the acetyl group, by binding them as a thioester. This coenzyme is required by a multitude of different reactions.

Pantothenic acid is found in many foods, and its deficiency is rare.

Lipoic Acid

Reduced lipoic acid is a short-chain fatty acid with two sulfhydryl groups. After oxidation, these groups form a disulfide linkage, as shown previously in the discussion of thiamine (TPP) metabolism. Although lipoic acid is itself insoluble in water, it forms water-soluble salts.

Two major reactions that utilize lipoic acid are the pyruvate and α-ketoglutarate dehydrogenase reactions, each of which utilizes TPP, lipoic acid, CoA—SH, FAD, and NAD$^+$.

No lipoic-acid deficiency state has been seen yet in humans.

Pyridoxal Phosphate

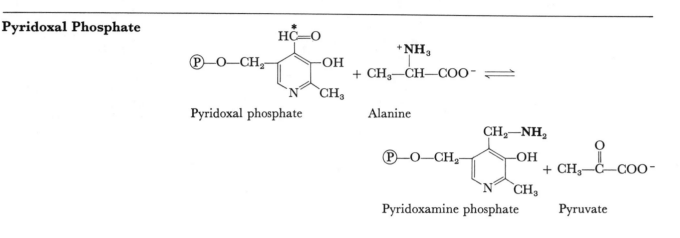

Pyridoxal phosphate   Alanine

Pyridoxamine phosphate   Pyruvate

Vitamin B$_6$ includes pyridoxal phosphate, pyridoxamine phosphate, and pyridoxine, the latter two of which humans convert to pyridoxal phosphate. Like nicotinamide, vitamin B$_6$ contains a pyridine ring.

Pyridoxal phosphate is required to transfer amino groups in transamination reactions and to decarboxylate amino acids. It is the "claw" of amino-acid metabolism, in that it brings amino acids into contact with the enzymes that metabolize them.

As shown above, pyridoxal phosphate removes the α-amino group of alanine to produce pyruvate, the corresponding α-ketoacid. The pyridoxamine phosphate generated can then donate its amine group to another α-ketoacid to transform it into an amino acid.

Vitamin B$_6$ deficiency is found in most pregnant women and in approximately 40% of alcoholics. It is also seen after chronic administration of B$_6$ antagonists such as isoniazid and penicillamine. Its features include hypochromic anemia, peripheral neuropathy, irritability, convulsions, and glossitis. Vitamin B$_6$ deficiency can lead to niacin deficiency, because B$_6$ is required to convert tryptophan to niacin.

Vitamin B$_6$ is required in proportion to the protein intake. It may be obtained from a variety of foods.

Biotin

Man acquires biotin, a B vitamin, both from the diet and from intestinal bacteria. Because of this dual supply, its deficiency is rare. Excessive ingestion of raw egg whites causes biotin deficiency, because avidin, a protein present in raw egg whites, binds to biotin and prevents its absorption. Cooking the eggs denatures avidin.

Biotin is bound to the ε-amino group of the lysine of a carboxylase enzyme. The essential role of biotin is to perform carboxylation: N-carboxybiotin donates its COO$^-$ group to a substrate to regenerate biotin.

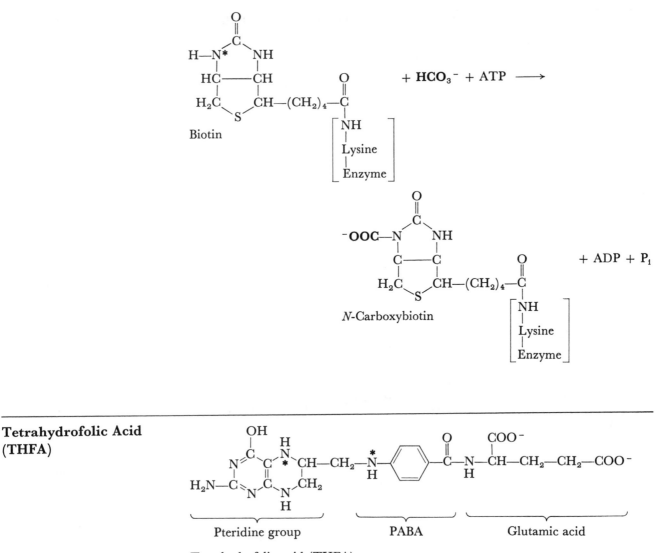

Tetrahydrofolic Acid (THFA)

Tetrahydrofolic acid (THFA)

Pteridine group   PABA   Glutamic acid

Folic acid, or pteroylglutamic acid, is a B vitamin that contains a pteridine ring, *p*-aminobenzoic acid (PABA), and glutamate. The principal folate in foods is called pteroylpolyglutamic acid because it contains a chain of glutamate residues that must be deconjugated (cleaved) before absorption. Alcoholics with cirrhosis cannot deconjugate these polyglutamates, and therefore they often develop folate deficiency. After absorption, humans reduce pteroylglutamic acid to THFA, whose structure is shown above.

The coenzymatic role of both THFA and vitamin B_{12} (cyanocobalamin) is to carry one-carbon groups such as methyl (CH_3), methylene (CH_2), formyl (CHO), and formimino (CH=NH) groups. This one-carbon pool *does not* include carboxyl groups, which are removed by TPP or pyridoxal phosphate and are added by biotin. Not only does THFA transfer one-carbon groups, it also oxidizes or reduces them. Such one-carbon transfers occur in various pathways, such as in de novo purine synthesis, the conversion of serine to glycine, and the methylation of deoxyuridylic (dUMP) to deoxythymidylic acid (dTMP), an essential step in DNA synthesis:

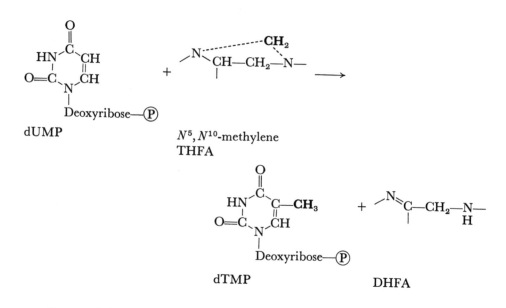

N^5, N^{10}-methylene
THFA

Folate deficiency causes macrocytic, megaloblastic anemia, because it slows both de novo purine synthesis and the conversion of dUMP to dTMP, thereby retarding DNA synthesis.

Foods rich in folic acid include green leafy vegetables, certain fresh fruits, and liver.

Cyanocobalamin

The intestinal absorption of cyanocobalamin, or vitamin B_{12}, depends on the gastric secretion of a glycoprotein, termed *intrinsic factor*, which combines with B_{12} and facilitates its absorption in the distal ileum. Without intrinsic factor, very little dietary B_{12} can be absorbed and the vitamin B_{12} excreted into the bile cannot be reabsorbed. The liver stores relatively large amounts of this vitamin. Even when vitamin B_{12} malabsorption occurs, symptoms of deficiency may not develop until the lapse of months or years.

Only microorganisms can synthesize vitamin B_{12}. Except for legume nodules, which harbor the B_{12}-producing bacteria, plants lack B_{12}. Animals obtain the vitamin from microorganisms and from eating other animals. A strict vegetarian diet, which excludes milk and eggs, has virtually no vitamin B_{12}. Strict vegetarians in India, however, may continue for decades without developing vitamin B_{12} deficiency, because, with a normal stomach and terminal ileum, they reabsorb most of the B_{12} they excrete into the bile.

After cyanocobalamin is absorbed, the cyanide is removed, and it is converted to the two active cobamide coenzymes: methylcobalamin (methyl-B_{12}) and deoxyadenosylcobalamin (DA-B_{12}), which is shown in Figure 5-1. Both coenzymes transfer one-carbon groups, such as methyl groups. Methyl-malonyl-CoA mutase requires DA-B_{12} to change the position of its (C=O)CoA to create succinyl-CoA (see Ch. 4, Problem 2).

Vitamin B_{12} deficiency can produce megaloblastic anemia, leukopenia, and thrombocytopenia, which closely resemble the findings in folate deficiency. This similarity may occur because vitamin B_{12} helps convert methyl-THFA

Figure 5-1

to THFA. Without sufficient B$_{12}$, methyl-THFA builds up, and too little methylene-THFA remains to convert dUMP to dTMP.

Unlike folate deficiency, vitamin B$_{12}$ deficiency impairs myelin formation by an unknown mechanism, and it can produce subacute combined degeneration of the spinal cord.

Ascorbic Acid

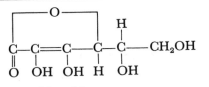

Ascorbic acid

Most species of animals synthesize their own ascorbic acid, or vitamin C, from related hexose sugars. Man, other primates, and guinea pigs, however, cannot do so, because they lack several enzymes that are necessary to create the essential —C(OH)=C(OH)— of ascorbate.

Ascorbic acid functions both as an antioxidant and as a coenzyme utilized in oxygenation reactions. Its roles include the oxygenation of proline and lysine (hydroxyproline and hydroxylysine are needed for synthesizing collagen) and the oxygenation of certain aromatic compounds. Ascorbate also improves duodenal iron absorption.

The features of scurvy (severe vitamin C deficiency) include follicular hyperkeratosis, petechiae, subconjunctival hemorrhage, gum changes, subperiosteal hemorrhage, and, in children, a failure to grow.

The major food sources of vitamin C are fresh fruit and vegetables. Of the meats eaten by Americans, only liver has appreciable quantities of this vitamin. Pasteurized milk and grain contain virtually no ascorbate. In the American diet, the main source of vitamin C is the potato, primarily because it is eaten in large amounts. Potatoes have as much ascorbate per gram as tangerines and grapefruit but less than half as much as oranges.

Baker et al. (1969) have shown that the total body pool of ascorbate in adults is 1.5 g. People who give themselves greatly excessive doses of ascorbate daily (such as 0.5 to 3.0 g) spill almost all of this dose into the urine once they have saturated their body pool with 1.5 g of ascorbate.

COENZYMES NOT CONTAINING VITAMINS

Not all coenzymes contain vitamins, as exemplified by coenzyme Q (the ubiquinones).

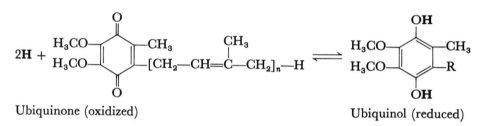

Ubiquinone (oxidized)　　　　　　　　　　　　　　Ubiquinol (reduced)

Coenzyme Q (CoQ) transfers H atoms and electrons in the mitochondrial oxidative-phosphorylation system. Humans synthesize their own ubiquinones.

VITAMINS NOT ACTING AS COENZYMES—THE FAT-SOLUBLE VITAMINS

Vitamin A

Humans ingest two types of vitamin A: provitamin A from plants and preformed vitamin A from animal sources.

β-Carotene, the principal dietary provitamin A, abounds in yellow and orange vegetables. As shown below, each molecule of β-carotene is cleaved in the intestinal mucosa to two molecules of retinol, or vitamin A_1.

The main sources of pre-formed vitamin A, or retinol, are liver, whole milk, fish oils, and eggs. Muscle meats and plants do not contain pre-formed vitamin A.

Vitamin A_1 is oxidized to retinal, or vitamin-A aldehyde, which combines with opsin, a protein, to form rhodopsin, the light-sensing pigment in the retina. Thus, the earliest symptom of vitamin A deficiency is night blindness.

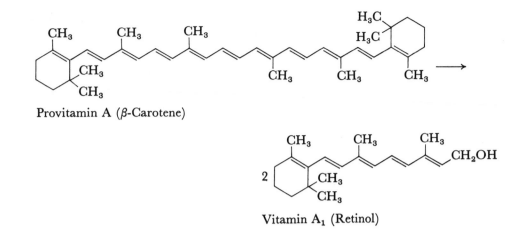

Provitamin A (β-Carotene)

Vitamin A_1 (Retinol)

In addition, vitamin A is required to form and maintain epithelial surfaces through a mechanism that is still unknown. Vitamin A deficiency causes follicular hyperkeratosis (i.e., the development of keratin plugs in hair follicles, as seen in scurvy) and a xerophthalmia (i.e., corneal dryness) that can progress to corneal ulcers and resultant blindness.

Acute vitamin A intoxication has occurred in Arctic explorers who ate polar bear livers. Chronic hypervitaminosis A usually occurs after enormously excessive vitamin A ingestion by food faddists or in the treatment of acne. Its features may include arthralgias, fatigue, night sweats, and headaches due to benign intracranial hypertension (pseudotumor cerebri).

Though harmless, excessive β-carotene ingestion makes the skin yellow or orange. In distinction to the observation in cases of jaundice, the sclera remains white.

Vitamin D

Vitamin D is called the "solar vitamin," because its synthesis involves the ultraviolet irradiation of sterols, such as 7-dehydrocholesterol, in human skin (shown below) or of ergosterol from milk.

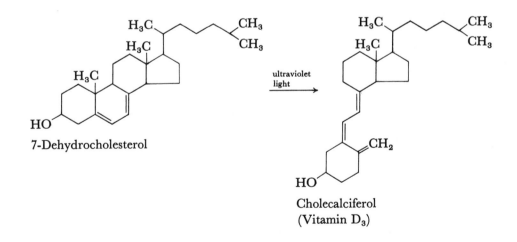

7-Dehydrocholesterol

ultraviolet light

Cholecalciferol
(Vitamin D_3)

The only foods that naturally contain vitamin D are fresh oils, such as fish liver oils. Milk becomes a source of vitamin D only after irradiating it or adding vitamin D. Humans exposed to bright sunlight year-round do not require dietary vitamin D.

Cholecalciferol, or vitamin D_3, must be hydroxylated twice to become metabolically active. It is hydroxylated first in the liver to produce 25-hydroxy-cholecalciferol (25-HCC), and next in the kidney to generate 1,25-dihydroxy-cholecalciferol (1,25-DHCC), the most active form of vitamin D. 1,25-DHCC acts as a hormone. Its principal role is to facilitate intestinal calcium absorption. In addition, it aids parathyroid hormone in mobilizing bone calcium.

The most common abnormality of vitamin D metabolism in America is due to the inability of the kidney to hydroxylate 25-HCC to 1,25-DHCC, which is usually the result of chronic renal failure.

Vitamin D deficiency in children, known as rickets, is most common in areas lacking sunshine. Rickets deforms the growing bones of children far more than does vitamin D deficiency in adults, which is termed osteomalacia.

Hypervitaminosis D occurs only after chronic, greatly excessive vitamin D administration. It produces hypercalcemia with resultant metastatic calcification and kidney stone formation.

Vitamin E

Vitamin E consists of a group of tocopherols, the α-tocopherols being the most active.

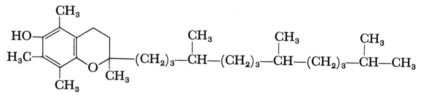

α-Tocopherol

The main dietary source of vitamin E is vegetable oils. It also occurs in grains and leafy vegetables.

The only established role of vitamin E in humans is to protect the poly-unsaturated fats and vitamin A from oxidation. As an antioxidant, it also protects the erythrocytes.

Deficiency of vitamin E (and other fat soluble vitamins) is common in people with fat malabsorption syndromes, and leads to hemolysis (rapid destruction of erythrocytes).

Vitamin K

Vitamin K is a group of napthoquinones with long, branched hydrocarbon side chains.

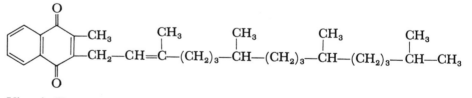

Vitamin K_1

The main dietary source of vitamin K is from green leafy vegetables. In addition, intestinal bacteria synthesize this vitamin.

Vitamin K acts in the liver to promote the synthesis of prothrombin and coagulation factors VII, IX, and X. Deficiency will reduce the plasma concentrations of these clotting factors and predispose to hemorrhage. Vitamin K deficiency can occur during the first few days after birth, because newborns lack the intestinal bacteria that produce vitamin K and because they have no store of vitamin K (it does not cross the placenta). Hence, all newborns are given a vitamin K injection to prevent hemorrhagic disease. Vitamin K deficiency may also occur following antibiotic therapy that sterilizes the gut.

Dicumarol drugs are important oral anticoagulants that act by antagonizing the action of vitamin K in the liver.

FOOD GROUPS

The most useful classification of foods is provided by the four basic food groups. A diet containing the recommended intake of these four groups is likely to be adequate. Nutritional deficiencies and imbalances are apt to occur if any one group is excluded or underrepresented in the diet without thoughtful planning to compensate for its absence.

In addition to these four food groups, there are accessory foods that add calories, such as butter, vegetable oils, sugar, and alcohol.

Meat Group

This group includes legumes (especially dried beans, peas, and nuts) as well as eggs, fish, meat, and poultry. These foods are characterized by their high concentration of protein of high *biologic quality* (which is explained in Chapter 13). They provide a good source of thiamine, niacin, cyanocobalamin, and iron. Liver and eggs abound in pre-formed vitamin A.

Milk Group

These foods provide the best sources of calcium and riboflavin. They are good sources of pre-formed vitamin A, vitamin D (in fortified milk), and cyanocobalamin. They have a moderate concentration of protein of good biologic quality. Cheeses, however, have a high protein concentration.

Vegetable-Fruit Group

This group is extremely low in protein and fat. The best sources of provitamin A are the yellow and orange vegetables. This is virtually the only food group that supplies ascorbate. Green leafy vegetables are rich in folic acid, vitamin E, and vitamin K.

Cereal Group

This group includes wheat products, such as bread and noodles, as well as rice, corn, oats, and other grains. Whole grains are good sources of thiamine. In addition, whole wheat is rich in iron. Processed grains are enriched to make them good sources of niacin, riboflavin, and iron. The grains have a moderate protein concentration of medium to low biologic quality.

Problem 1

Which coenzyme is *not* derived from a vitamin?

A. CoA—SH
B. FAD^+
C. NAD^+
D. Pyridoxal phosphate
E. CoQ
F. TPP

Problem 2

Intestinal bacteria supply significant quantities of which two human vitamins?

A. Riboflavin, niacin
B. B_{12}, vitamin D
C. Biotin, vitamin K
D. Ascorbate, thiamine

Problems 3–11

Match the vitamins below to their descriptions:

A. Pyridoxal phosphate
B. Cyanocobalamin
C. Niacin
D. Pantothenic acid
E. Thiamine
F. Riboflavin
G. Biotin
H. Pteroylpolyglutamic acid

3. Alcoholics with cirrhosis cannot deconjugate this vitamin in their intestine; hence, it is poorly absorbed, and they can develop macrocytic anemia.
4. Deficiency disease is widespread in populations living mainly on polished, milled rice.
5. It is partially synthesized from dietary tryptophan.
6. It is a component of FAD.
7. Its deficiency causes subacute, combined degeneration of the spinal cord along with megaloblastic, macrocytic anemia.
8. It is involved in the transamination and decarboxylation of amino acids.
9. It is a component of CoA.
10. This may be required in the carboxylation of acetate to malonate.
11. Its deficiency in alcoholics causes high-output congestive heart failure.

Problem 12

Which statement about the RDA is *incorrect*?

A. Meeting the RDA standards insures adequate vitamin intake for most healthy Americans.
B. The RDA always exceeds the MDR.
C. Meeting the RDA standards insures adequate vitamin intake for most chronically ill patients.
D. The RDA standards are population estimates.

Problem 13

Which vitamin is utilized to transfer methyl and formyl groups?

A. Thiamine
B. Ascorbic acid
C. Folic acid
D. CoA
E. Riboflavin

Problem 14

Which vitamin deficiency would most likely occur in a patient with malabsorption syndrome due to the inadequate secretion of pancreatic enzymes found in cystic fibrosis?

A. Niacin
B. Folic acid
C. Vitamin A
D. Pantothenic acid
E. Pyridoxamine phosphate

Problem 15

Which chemical compound below is *incorrectly* labeled?

A. α-Tocopherol, vitamin E
B. Thiamine, vitamin B_1
C. Ascorbic acid, vitamin C
D. Cyanocobalamin, vitamin B_{12}
E. Pyridoxal phosphate, vitamin B_2

Problem 16

A patient who excludes the vegetable-fruit food group from his diet will most likely develop a deficiency of which two of the nutrients below?

A. Vitamin A
B. Essential fatty acids
C. Ascorbate
D. Thiamine
E. Niacin
F. Cyanocobalamin

Problem 17

Which statement about vitamin D is *incorrect*?

A. It is converted into a hormone.
B. Children in the tropics usually do not require dietary vitamin D.
C. Raw milk is a good source of vitamin D.
D. 7-Dehydrocholesterol in human skin is converted to cholecalciferol by ultraviolet irradiation.

Problems 18–20

Match the structures below to the compounds named in Problems 18–20:

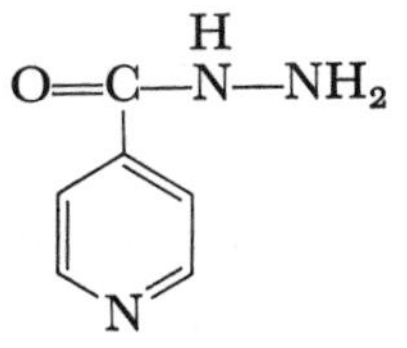
H
O=C—N—NH2
N

A

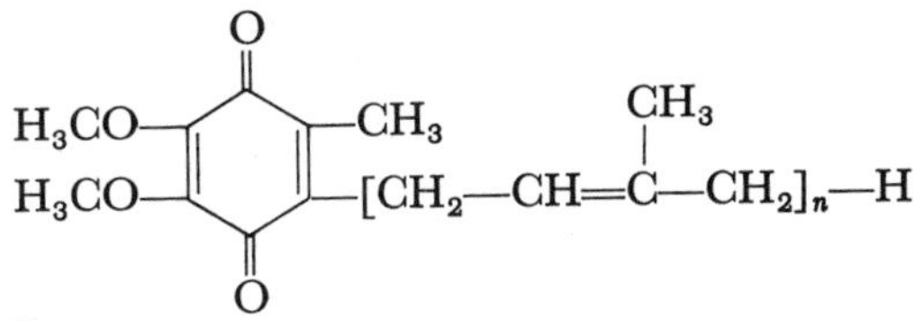
O
H3CO
H3CO
CH3
CH3
[CH2—CH=C—CH2]n—H
O

B

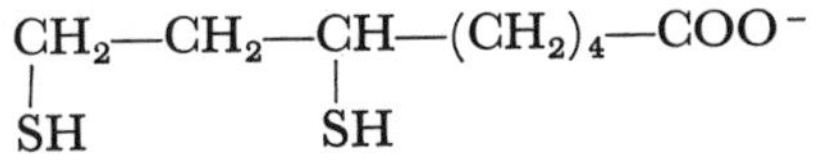
CH2—CH2—CH—(CH2)4—COO⁻
SH
SH

C

18. Lipoic acid.
19. Antagonist to pyridoxal phosphate and pyridoxamine phosphate.
20. CoQ (ubiquinone).

ANSWERS

1. E.
2. C.
3. H.
4. E.
5. C.

6. F.
7. B.
8. A.
9. D.
10. G.
11. E.
12. C. (The RDA standards were created for healthy Americans and do not necessarily apply to those with chronic disease. Diseases causing fat malabsorption, for example, impair the absorption of fat-soluble vitamins. Meeting the RDA in such patients will still result in vitamin deficiencies. These patients require injections of fat-soluble vitamins or the administration of special water-solubilized forms of these vitamins.)
13. C.
14. C.
15. E.
16. A and C.
17. C. (Raw milk is devoid of vitamin D. Irradiating milk, however, converts its ergosterol to vitamin D.)
18. C. (Lipoic acid can be easily recognized if you remember that it is a short-chain fatty acid with two sulfhydryl groups.)
19. A. (This is isoniazid, a drug used to treat tuberculosis. Like pyridoxamine phosphate, it contains an amine group bound to a pyridine nucleus.)
20. B. (The ubiquinones may be recognized by their benzoquinone nucleus.)

REFERENCES

Baker, E., Hodges, R. E., Hood, J., Sauberlich, H. E., and March, S. C. Metabolism of ascorbic-1-^{14}C acid in experimental human scurvy. *Am. J. Clin. Nutr.* 22:549, 1969.

Deluca, H. F. Metabolism of vitamin D: current status. *Am. J. Clin. Nutr.* 29:1258, 1976.

Farrel, P. M., Bieri, J. G., Fratantoni, J. F., Wood, R. E., and Di Sant'Agnese, P. A. The occurrence and effects of human vitamin E deficiency. A study in patients with cystic fibrosis. *J. Clin. Invest.* 60:233, 1977.

Friedman, P. J., and Hodges, R. E. Tongue colour and B-vitamin deficiencies. *Lancet* 1:1159, 1977.

Goodhart, R. S., and Shils, M. E. *Modern Nutrition in Health and Disease* (5th ed.). Philadelphia: Lea & Febiger, 1973. Pp. 142–258.

Goodman, L. S., and Gilman, A. *The Pharmacological Basis of Therapeutics* (5th ed.). New York: Macmillan, 1975. Pp. 1544–1560, 1564–1598.

Lehninger, A. L. *Biochemistry: The Molecular Basis of Cell Structure and Function* (2nd ed.). New York: Worth, 1975. Pp. 335–360.

Siegel, G. J., Albers, R. W., Katzman, R., and Agranoff, B. W. *Basic Neurochemistry* (2nd ed.). Boston: Little, Brown, 1976. Pp. 605–626 (Vitamin and Nutritional Deficiencies).

6 Energetics

The three fundamental thermodynamic variables are enthalpy (H), entropy (S), and free energy (G).

In biochemistry the standard state is defined as pH 7.0, 25°C (298°K), all solutes at 1 molar concentration, and all gases at 1 atm pressure. A superscript zero prime is used to denote standard state conditions in biochemistry, as in $G^{0'}$, $H^{0'}$, and $S^{0'}$. The inorganic standard states, designated by a superscript zero, differ in that the standard pH is 0.0. The actual conditions are indicated by G, H, and S.

ENTHALPY

Enthalpy, H, is defined as the heat content of a physical object or body; it is the sum of the internal energy (E) plus the pressure-volume product (PV):

$$H = E + PV \tag{6-1}$$

The *standard enthalpy of formation* of a chemical compound A, $\Delta H_F^{0'}(A)$, is the enthalpy change, or increase of heat content, due to the reaction that generates one mole of A from its constituent elements in their standard states, e.g., O_2, N_2, C (graphite), S, H_2, or whatever. The enthalpy of an element is arbitrarily taken to be zero. For the reaction below, the standard enthalpy change, $\Delta H^{0'}$, is calculated by subtracting the standard enthalpies of formation of the reactants from those of the products:

$$A + B \longrightarrow C + D$$

$$\Delta H^{0'} = \Delta H_F^{0'}(C) + \Delta H_F^{0'}(D) - \Delta H_F^{0'}(A) - \Delta H_F^{0'}(B) \tag{6-2}$$

ENTROPY

Entropy, S, is defined as the degree of randomness or disorder of a system. Carbon dioxide gas inside a bottle has a high entropy, because the molecules can move freely in any direction. The glass of the bottle, on the other hand, has a rigid crystalline structure that constrains freedom of molecular movement within that structure; hence, it has a low entropy. The entropy of a pure, crystalline solid at 0°K (absolute zero) is zero. The entropy increases with changes of phase; i.e., as the substance goes from solid to liquid to gas.

The *second law of thermodynamics* predicts that for a closed system, the entropy of a system and its surroundings will always increase. A closed system is one where no energy is being put in from an outside source. Unfortunately, the entropy change, ΔS, cannot be measured directly. Instead, ΔS is calculated from ΔH, ΔG, and ΔT, where ΔH is the enthalpy change, ΔG the free energy change, and ΔT the temperature change.

If $\Delta S < 0$, the randomness of the system declines; that is, the products have more order than the reactants. If $\Delta S > 0$, the products have a greater degree of disorder than the reactants.

FREE ENERGY

The *free energy*, G, is the maximum usable work that can be obtained from a system at constant pressure, temperature, and volume. The energy spent to maintain randomness, TS (where T is measured in degrees Kelvin), cannot be harnessed to perform work. Hence,

$$G = H - TS \tag{6-3}$$

$$\Delta G = \Delta H - T\Delta S \tag{6-4}$$

The *standard free energy of formation* of A, $\Delta G_F^{0\prime}(A)$, is defined as the change in free energy during the synthesis of one mole of A from its constituent elements under standard conditions.

The *standard free energy change* of a reaction, $\Delta G^{0\prime}$, can be calculated by subtracting the $\Delta G_F^{0\prime}$ values of the reactants from those of the products:

$$\Delta G^{0\prime} = \Delta G_F^{0\prime}(C) + \Delta G_F^{0\prime}(D) - \Delta G_F^{0\prime}(A) - \Delta G_F^{0\prime}(B) \tag{6-5}$$

The standard free energy of reaction, $\Delta G^{0\prime}$, can also be calculated from the *equilibrium constant* for the reaction, K_{eq}. The actual free energy change (ΔG) in calories per mole is related to $\Delta G^{0\prime}$:
where

$$K = \frac{[C] \times [D]}{[A] \times [B]}$$

$$\Delta G = \Delta G^{0\prime} + 4.57 T \log K \tag{6-6}$$

and T is in degrees Kelvin.

In the standard state, K equals 1.0, since by definition each reactant and product is at 1 molar concentration, and ΔG equals $\Delta G^{0\prime}$.

At equilibrium, K equals K_{eq} and ΔG equals zero (a system at equilibrium produces no usable work). Substituting into Equation 6-6 and rearranging:

$$\Delta G^{0\prime} = -4.57 T \log K_{eq} \tag{6-7}$$

Having calculated $\Delta G^{0\prime}$ in calories per mole from K_{eq} using Equation 6-7, one can now use Equation 6-6 to calculate ΔG for any actual temperature and reactant concentrations.

If $\Delta G < 0$, then the reaction will proceed spontaneously. Such a reaction is *exergonic*, because it releases energy that can perform work. Only if properly harnessed, however, will this energy perform actual work.

If $\Delta G > 0$, then the forward reaction will not proceed spontaneously. Outside energy must be added to drive the reaction forward from A and B to the products C and D. Such a reaction is *endergonic*, or energy-requiring. The reverse reaction—i.e., from C and D to A and B—will have a negative value of ΔG and will tend to proceed spontaneously.

If ΔG equals zero, then the reaction is at equilibrium, i.e., the rate of the forward reaction is equal to the rate of the reverse reaction. The concentrations of the reactants and products at this point define the equilibrium constant, K_{eq}, for the reaction. (These concentrations need not be 1 molar, so the equilibrium state is different from the standard state.)

The actual conditions of most reactions—i.e., temperature, pressure, or concentration—can be adjusted to make ΔG positive, zero, or negative. However, it must be kept in mind that ΔG *gives no information about the reaction rate; it only indicates the direction of spontaneous reaction.* Its negative value may predict a forward reaction, but the reaction may proceed very slowly.

Living beings require a constant energy supply from their surroundings to maintain their high degree of internal order. Most individual reactions inside a cell reach equilibrium only when the cell dies. During life, these reactions accomplish rapid metabolic interconversions and operate nowhere near their equilibrium points.

Problem 1

Calculate $\Delta H^{0\prime}$, $\Delta G^{0\prime}$, and $\Delta S^{0\prime}$ for the hypothetical reaction of A and B to yield C. At standard state, is this reaction endergonic or exergonic? Will it proceed spontaneously? Does the state of the product become more or less random?

$$A + B \rightleftharpoons C$$

$K_{eq} = 3.16 \times 10^{-13}$, $T = 298°K$, $\Delta H_F^{0\prime}(A) = -20.0$ kcal/mole, $\Delta H_F^{0\prime}(B) = -10.0$ kcal/mole, $\Delta H_F^{0\prime}(C) = -80.0$ kcal/mole.

Problem 2

Using the value of $\Delta G^{0\prime}$ from Problem 1, calculate ΔG at 37°C (310°K) for the following reactant concentrations. Decide whether each reaction is endergonic or exergonic and whether it will occur spontaneously.

	[A]	[B]	[C]
A.	1.00×10^{-3}	2.00×10^{-5}	2.00×10^{-6}
B.	0.100	0.30	6.00×10^{-15}
C.	1.00×10^{-2}	1.00×10^{-2}	1.00×10^{-16}

HIGH- AND LOW-ENERGY PHOSPHATE COMPOUNDS

Bond energy, to an inorganic chemist, is the energy needed to break a bond. When biochemists refer to high- and low-energy phosphate bonds, however, they are talking about the $\Delta G^{0\prime}$ for the reaction that hydrolyzes the bond, instead of the energy required to break the bond itself.

Glycerol-1-phosphate has a low-energy phosphate bond, because $\Delta G^{0\prime}$ for phosphate hydrolysis is relatively low (-2.3 kcal/mole).

Glycerol-1-P   Glycerol

Adenosine triphosphate (ATP) is the principal intracellular energy currency. It has an intermediate value of $\Delta G^{0\prime}$ for its hydrolysis (-7.3 kcal/mole), but it is often called a high-energy phosphate compound. Adenosine diphosphate (ADP) has the same intermediate value of $\Delta G^{0\prime}$ for hydrolysis (-7.3 kcal/mole). Adenosine monophosphate (AMP) differs in that it is a low-energy phosphate compound ($\Delta G^{0\prime} = -3.4$ kcal/mole).

$ATP + H_2O \rightleftharpoons ADP + P_i$	$\Delta G^{0\prime} = -7.3$ kcal/mole
$ADP + H_2O \rightleftharpoons AMP + P_i$	$\Delta G^{0\prime} = -7.3$ kcal/mole
$AMP + H_2O \rightleftharpoons adenosine + P_i$	$\Delta G^{0\prime} = -3.4$ kcal/mole

Pyrophosphate hydrolysis of ATP yields more energy than phosphate removal. The 10.0 kcal/mole released drives reactions that demand more than the 7.3 kcal/mole liberated during ATP hydrolysis to ADP and P_i.

$$ATP + H_2O \rightleftharpoons AMP + PP_i \qquad \Delta G^{0\prime} = -10.0 \text{ kcal/mole}$$

As a further "push" to certain reactions, a pyrophosphatase cleaves PP_i to release 4.6 kcal/mole.

$$PP_i + H_2O \rightleftharpoons 2P_i \qquad \Delta G^{0\prime} = -4.6 \text{ kcal/mole}$$

Creatine phosphate, the principal energy reservoir in muscle, has a phosphate bond of higher energy than that of ATP, with a $\Delta G^{0\prime}$ of -10.3 kcal/mole. In cells, the hydrolysis of such high-energy phosphate compounds is coupled to the phosphorylation of ADP to ATP. As shown below, creatine phosphate donates its phosphate to ADP to create ATP:

Creatine phosphate $+ H_2O \longrightarrow$ creatine $+ P_i$	$\Delta G^{0\prime} = -10.3$ kcal/mole
$ADP + P_i \longrightarrow ATP + H_2O$	$\Delta G^{0\prime} = +7.3$ kcal/mole

Net: Creatine phosphate $+ ADP \longrightarrow$ creatine $+ ATP$ $\qquad \Delta G^{0\prime} = -3.0$ kcal/mole

Other major high-energy phosphate compounds include phosphoenolpyruvate and 1,3-diphosphoglycerate. The hydrolysis of phosphate from each is coupled to ATP formation in glycolysis.

In the cell, endergonic reactions can be driven by coupling them to an exergonic reaction, such as ATP hydrolysis. The phosphorylation of glucose with P_i to glucose-6-phosphate, for instance, has a positive $\Delta G^{0\prime}$ of 3.3 kcal/mole. ATP donates the P_i to glucose and simultaneously supplies the energy to drive this reaction toward glucose-6-phosphate formation:

Glucose $+ P_i \longrightarrow$ glucose-6-P $+ H_2O$	$\Delta G^{0\prime} = +3.3$ kcal/mole
$ATP + H_2O \longrightarrow ADP + P_i$	$\Delta G^{0\prime} = -7.3$ kcal/mole

Net: Glucose $+ ATP \longrightarrow$ glucose-6-P $+ ADP$ $\qquad \Delta G^{0\prime} = -4.0$ kcal/mole

Calculate $\Delta G^{0\prime}$ for the pyruvate kinase reaction below. The $\Delta G^{0\prime}$ values for the hydrolysis of phosphate from phosphoenolpyruvate (PEP) and ATP are -14.8 and -7.3 kcal/mole, respectively.

$$CH_2\!\!=\!\!\overset{\overset{\displaystyle O-\textcircled{P}}{|}}{C}\!\!-\!\!COO^- + ADP \longrightarrow CH_3\!\!-\!\!\overset{\overset{\displaystyle O}{\|}}{C}\!\!-\!\!COO^- + ATP$$

PEP   Pyruvate

OXIDATION-REDUCTION REACTIONS

Oxidation is defined as loss of electrons, while *reduction* is a gain of electrons; Fe^{+2}, for example, is more reduced than Fe^{+3}, whereas Cu^{+2}, on the other hand, is more oxidized than Cu^+.

The *oxidation state* of a carbon atom depends on the electronegativity of the atoms bound to it. The carbon atom of methane (CH_4) represents the state of greatest reduction, because carbon and hydrogen equally share electrons. The carbon atom of methanol (CH_3OH) is more oxidized, because the hydroxyl oxygen is more electronegative than carbon. The C—OH bond is mildly polar; the carbon has a slight positive charge and has partially given up an electron to oxygen. The carbon atom of formaldehyde (HCHO) is even more oxidized; the carbonyl bond is more polarized than the C—OH bond of methanol. The carbon atom of formic acid (HCOOH) is more oxidized than that of formaldehyde. Carbon dioxide (O=C=O) has, in turn, a more oxidized carbon atom than does formic acid.

The *reduction potential*, $E^{0\prime}$, is the electrical potential (E) in volts (V) measured during the reduction reaction under standard conditions. The $H_2:H^+$ electrode is used as the reference or zero potential:

$$\tfrac{1}{2}H_2 \longrightarrow H^+ + e^- \qquad E^{0\prime} = 0.00 \text{ V}$$

A compound that can be oxidized more readily than H_2 will have a negative $E^{0\prime}$. The *oxidation potential* of a reaction has the same absolute value as the reduction potential but the opposite sign.

An *oxidation-reduction reaction* involves a transfer of electrons among its reactants. Each overall oxidation-reduction reaction consists of two *half-reactions*. The mitochondrial electron-transport system, for example, couples the oxidation of NADH to NAD^+ and H^- (this hydride ion represents $H^+ + 2e^-$) with the formation of water from $2H^+$, $\tfrac{1}{2}O_2$, and $2e^-$. The standard reduction potential of NAD^+ to NADH is -0.32 V. Therefore, the oxidation of NADH has an $E^{0\prime}$ of $+0.32$ V. The standard reduction potential for water formation from O_2, H^+, and e^- is $+0.816$ V. The difference between the standard reduction potentials, $\Delta E^{0\prime}$, is the *net reaction potential* under standard conditions, i.e., $25°C$, all solutes at 1 molar concentration, and all gases at 1 atm pressure. To calculate $\Delta E^{0\prime}$, the oxidation potential of the oxidation half-reaction is added to the reduction potential of the reduction half-reaction:

$$NADH \longrightarrow NAD^+ + H^+ + 2e^- \qquad\qquad E = -E^{0\prime} = +0.320 \text{ V}$$
$$\tfrac{1}{2}O_2 + 2H^+ + 2e^- \longrightarrow H_2O \qquad\qquad E = E^{0\prime} = +0.816 \text{ V}$$

Net: $NADH + \tfrac{1}{2}O_2 + H^+ \longrightarrow NAD^+ + H_2O \qquad \Delta E^{0\prime} = +1.14 \text{ V}$

When ΔE, the reaction potential under actual (not necessarily standard) conditions, is positive, a reaction will proceed spontaneously. When ΔE equals zero, the reaction is at equilibrium.

Knowing $\Delta E^{0\prime}$ and n, the number of electrons transferred in the reaction, one can calculate $\Delta G^{0\prime}$ in kilocalories per mole:

$$\Delta G^{0\prime} = -23.1(n)(\Delta E^{0\prime}) \tag{6-8}$$

Problem 4	Calculate $\Delta G^{0\prime}$ for the mitochondrial electron-transport scheme just outlined in the text. Under standard conditions, is it endergonic or exergonic? Will it proceed spontaneously?

Problem 5

Chronic alcoholics require more ethanol than nondrinkers to become intoxicated, because their liver has more alcohol dehydrogenase to oxidize ethanol to acetaldehyde (CH_3CHO). Calculate $\Delta E^{0\prime}$ and $\Delta G^{0\prime}$ for the alcohol dehydrogenase reaction, given that $E^{0\prime}$ for alcohol dehydrogenase is -0.197 V.

$$NAD^+ + H^+ + 2e^- \longrightarrow NADH \qquad E = E^{0\prime} = -0.320 \text{ V}$$
$$CH_3CH_2OH \longrightarrow CH_3CHO + 2H^+ + 2e^- \qquad E = -E^{0\prime} + 0.197 \text{ V}$$

Net: $CH_3CH_2OH + NAD^+ \longrightarrow CH_3CHO + NADH + H^+$

ACTIVATION ENERGY

Activation energy, E_a, is the free energy needed to convert the reactants, such as the substrates of an enzymatic reaction, into their reactive states; once converted to the reactive or transition state, they are rapidly converted to products. The *reaction rate* is proportional to the quantity of reactants in the transition state. The higher the E_a, the slower the rate of reaction, because fewer of the reactant molecules will possess enough kinetic energy to become transformed into their reactive states.

Figure 6-1

Effect of catalysts on activation energy (E_a).

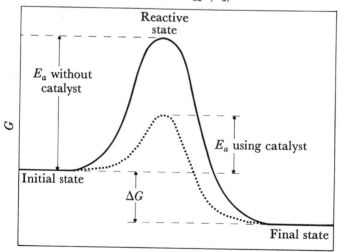

Progress of Reaction

One method for increasing the rate of an inorganic reaction is to raise the temperature, thereby promoting more of the reactants to the transition state. Another method is to add inorganic catalysts, such as platinum, which augment both the forward and reverse reaction rates by lowering the E_a barrier, as shown in Figure 6-1.

Since mammals cannot significantly raise their body temperature, they must rely on enzymes as catalysts to lower the E_a values for many reactions. Without enzymes, life could not continue, because its reactions would proceed too slowly.

Problem 6

To test your knowledge of nutrition, your 400-pound patient asks you to predict how many Calories (one food Calorie equals 1000 calories) there are in her bedtime snack of one cup of whole milk (9 g protein, 8 g triglyceride, and 12 g carbohydrate) and one piece of apple pie (3 g protein, 14 g triglyceride, and 51 g carbohydrate). Naturally, you recall that carbohydrate and protein each contain 4 Cal/gram, while triglyceride has 9 Cal/gram.

Problem 7

Calculate $\Delta G^{0\prime}$ for the malate dehydrogenase reaction below:

$$\overset{\displaystyle OH}{\underset{\displaystyle }{\text{-OOC—CH—CH}_2\text{—COO}^-}} + NAD^+ \rightleftharpoons$$
Malate

$$\overset{\displaystyle O}{\underset{\displaystyle }{\text{-OOC—C—CH}_2\text{—COO}^-}} + NADH + H^+$$
Oxaloacetate

Given below are the standard reduction potentials for the half-reactions:

$$\text{Oxaloacetate} + 2H^+ + 2e^- \longrightarrow \text{malate} \qquad E^{0\prime} = -0.166 \text{ V}$$
$$NAD^+ + H^+ + 2e^- \longrightarrow NADH \qquad E^{0\prime} = -0.320 \text{ V}$$

Problem 8

For reaction X, $\Delta G = -10$ kcal/mole, while for reaction Y, $\Delta G = -5$ kcal/mole. Choose the correct statement about their reaction rates:

A. Rate of reaction X will exceed the rate of Y.
B. Rate of reaction Y will exceed the rate of X.
C. If E_a of reaction Y is less than the E_a of X, then the rate of reaction Y will exceed the rate of X.
D. No valid comparisons of reaction rates can be made from these data.

Problem 9

Choose which endergonic reactions could be driven by coupling them to the hydrolysis of ATP to ADP + P_i:

A. Creatine + P_1 $\longrightarrow$
   creatine phosphate + H_2O $\qquad \Delta G^{0\prime} = +10.3$ kcal/mole

B. Glycerol + P_i $\longrightarrow$

        glycerol-1-phosphate + H_2O $\Delta G^{0\prime} = +2.2$ kcal/mole

C. Fructose-6-phosphate + P_i $\longrightarrow$

        fructose-1,6-diphosphate + H_2O $\Delta G^{0\prime} = +4.0$ kcal/mole

D. Pyruvate + P_i $\longrightarrow$

        phosphoenolpyruvate + H_2O $\Delta G^{0\prime} = +14.8$ kcal/mole

---

ANSWERS

1. $\Delta H = \Delta H_F{}^{0\prime}(C) - \Delta H_F{}^{0\prime}(A) - \Delta H_F{}^{0\prime}(B)$

     $= -80 - (-20) - (-10) = -50$ kcal/mole

$\Delta G^{0\prime} = -4.57\,T \log K_{eq}$

     $= -4.57(298°K) \log(3.16 \times 10^{-13})$ cal/mole

     $= 17.0 \times 10^3$ cal/mole or 17.0 kcal/mole

Thus, at standard state the reaction is endergonic and will not proceed spontaneously.

On substitution of the standard-state values, Equation 6-4 may be used to calculate $\Delta S^{0\prime}$:

$\Delta G^{0\prime} = \Delta H^{0\prime} - T\Delta S^{0\prime}$
17 kcal/mole $= -50$ kcal/mole $- (298°K)\Delta S^{0\prime}$
$\Delta S^{0\prime} = -0.225$ kcal/mole per degree

Hence, the state of the product is less random than that of the reactants.

2.   A. $K = \dfrac{[C]}{[A] \times [B]} = \dfrac{2.0 \times 10^{-6}}{(1.0 \times 10^{-3})(2.0 \times 10^{-5})} = 100$

     $\Delta G = \Delta G^{0\prime} + 4.57\,T \log K$

         $= 17.0 \times 10^3$ cal/mole $+ 4.57$ cal/mole/deg $(310°K) \log(100)$

         $= 19.8 \times 10^3$ cal/mole

Endergonic, not spontaneous.

B. $K = 2.00 \times 10^{-13}$

     $\Delta G = 17.0 \times 10^3$ cal/mole $+ 4.57(310) \log(2.00 \times 10^{-13})$ cal/mole

         $= (17.0 \times 10^3)$ cal/mole $- (18.0 \times 10^3)$ cal/mole

         $= -1.0 \times 10^3$ cal/mole

Though extremely low, the negative value of ΔG makes the forward reaction exergonic and spontaneous.

C. $K = 1.00 \times 10^{-12}$

     $\Delta G = (17.0 \times 10^3)$ cal/mole $+ 4.57(310) \log(1.00 \times 10^{-12})$ cal/mole

         $= (17.0 \times 10^3) - (17.0 \times 10^3) = 0$

Hence, the reaction is at equilibrium; the forward reaction rate equals the reverse rate.

3. PEP + H_2O $\longrightarrow$ pyruvate + P_1 $\Delta G^{0\prime} = -14.8$ kcal/mole
    ADP + P_1 $\longrightarrow$ ATP + H_2O $\Delta G^{0\prime} = +7.3$ kcal/mole

    ___

    Net: PEP + ADP $\longrightarrow$ pyruvate + ATP $\Delta G^{0\prime} = -7.5$ kcal/mole

4. Since $n = 2$ and $\Delta E^{0\prime} = +1.14$ V,

    $$\Delta G^{0\prime} = -23.1(n)(\Delta E^{0\prime})$$
    $$= -23.1 \times 2 \times 1.14$$
    $$= -52.7 \text{ kcal/mole}$$

    The reaction is exergonic and spontaneous under standard conditions.

5. $\Delta E^{0\prime} = -0.320 - (-0.197) = -0.320 + 0.197 = -0.123$ V
    $$\Delta G^{0\prime} = -23.1(n)(\Delta E^{0\prime})$$
    $$= -23.1 \times 2 \times (-0.123)$$
    $$= +5.68 \text{ kcal/mole}$$

6. Let CHO represent carbohydrate and TG triglyceride:
    Cal in whole milk = 4 Cal/gram × (9 g protein + 12 g CHO) + 9 Cal/gram × 8 g TG = 156 Cal
    Cal in apple pie = 4 Cal/gram × (3 g protein + 51 g CHO) + 9 Cal/gram × 14 g TG = 342 Cal
    Total "snack" Calories = 498

7. Rearranging these half-reactions gives:

    Malate $\longrightarrow$ oxaloacetate + $2H^+$ + $2e^-$ $E = -E^{0\prime} = +0.166$ V
    NAD^+ + H^+ + $2e^-$ $\longrightarrow$ NADH $E = E^{0\prime} = -0.320$ V

    ___

    Net: Malate + NAD^+ $\longrightarrow$
             oxaloacetate + NADH + H^+ $\Delta E^{0\prime} = -0.154$ V

    $$\Delta G^{0\prime} = -23.1(n)(\Delta E^{0\prime})$$
    $$= -23.1 \times 2 \times (-0.154)$$
    $$= +7.11 \text{ kcal/mole}$$

8. D. (Although it is true that lowering E_a for a given reaction will increase its rate, it does not follow that the relative rates of two different reactions can be deduced by comparing their E_a values, nor do the ΔG values give any information about reaction rates.)

9. B, C. (ATP hydrolysis to ADP + P_i liberates 7.3 kcal/mole, enough to drive reactions B and C but not A and D.)

REFERENCES

Bhagavan, N. V. *Biochemistry—A Comprehensive Review*. Philadelphia: Lippincott, 1974. Pp. 66–79.

Lehninger, A. L. *Biochemistry: The Molecular Basis of Cell Structure and Function* (2nd ed.). New York: Worth, 1975. Pp. 188–189, 390–414, 477–480.

White, A., Handler, P., and Smith, E. L. *Principles of Biochemistry* (5th ed.). New York: McGraw-Hill, 1973. Pp. 210–213, 233–234, 311–331, 373–375.

7 Structure and Properties of Carbohydrates

Carbohydrates are defined as polyhydroxylated compounds with at least three carbon atoms that may or may not possess a carbonyl group. The formulas of many carbohydrates show one water molecule for every carbon; hence, the name "carbohydrate."

The aldehyde sugars are called *aldoses*, while those with a ketone group are *ketoses*. In designating their relative position, the carbon atoms are numbered so as to assign the lowest possible number to the carbonyl carbon.

Sugar alcohols contain a hydroxyl group in place of a carbonyl group. Mannitol, the sugar alcohol derived from mannose, is frequently used medically as an osmotic diuretic to reduce cerebral edema. Sorbitol, the sugar alcohol derived from glucose, often accumulates in the lenses of diabetics and produces cataracts.

Sugar acids, such as glucuronic and ascorbic acids, contain a carboxyl group.

Other groups that can substitute for a hydroxyl group include phosphate, sulfate, amino, and *N*-acetyl groups. The deoxy sugars of DNA lack a hydroxyl group at carbon 2.

TRIOSES

The three-carbon sugars, the *trioses*, are the smallest possible carbohydrates. Several trioses are shown below according to the Fischer method of representation. If the OH group on the penultimate (next to last) carbon atom points to the right (*dextro*) on the Fischer structure, the sugar is the D isomer. If the OH group points to the left (*levo*), it is the L isomer of the sugar. These designations have nothing to do with optical isomerization—i.e., the ability to rotate the plane of polarized light—which is denoted by *d* or *l*.

Glyceraldehyde is an aldotriose, dihydroxyacetone (DHA) is a ketotriose, and glycerol is a sugar alcohol, as shown below. Lactic and pyruvic acids are derived from glyceraldehyde and DHA respectively; neither fits the definition of a carbohydrate, because each has only one hydroxyl group.

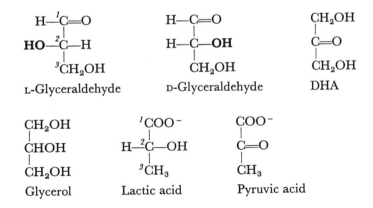

TETROSES AND PENTOSES

Four-carbon sugars, or *tetroses*, play a minor role in humans compared to the five-carbon (*pentose*) and six-carbon (*hexose*) sugars.

The major human pentoses are ribose and 2-deoxyribose. In solution, each exists primarily as a five-membered ring that contains oxygen and four carbons. Since this ring resembles that of furan, these sugars are said to be in furanose form.

The Haworth projections shown below place the furanose ring perpendicular to the plane of the page. If an OH group projects to the right on the Fischer structure, it will project downward on the Haworth projection.

The carbonyl carbon atom of pentoses and hexoses is termed the *anomeric* carbon atom. Ring formation links the anomeric carbon to the penultimate carbon and transforms it into an asymmetric carbon atom with two possible configurations. In the α-anomer, the OH group that is attached to this carbon atom is to the right on the Fischer structure and down on the Haworth projection. The β-anomer has the OH projecting to the left on the Fischer structure and upward on the Haworth projection. An asterisk denotes the anomeric carbon on the structures below.

The spontaneous interconversion of α and β anomers in solution is termed *mutarotation*.

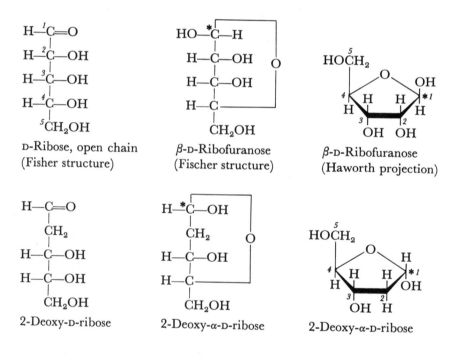

D-Ribose, open chain (Fisher structure)

β-D-Ribofuranose (Fischer structure)

β-D-Ribofuranose (Haworth projection)

2-Deoxy-D-ribose

2-Deoxy-α-D-ribose

2-Deoxy-α-D-ribose

HEXOSES

The two major *aldohexoses* in man, glucose and galactose, differ only in the configuration of their OH group at carbon 4; they are *epimers*. In solution, less than 1% of either remains as a straight chain. Instead, each forms a six-membered ring that contains five carbon atoms and one oxygen. Because this ring resembles that of pyran, these sugars are termed pyranoses.

The major *ketohexose* in humans is fructose. Fructose exists mainly in the furanose form with a small amount in the pyranose form, and it has carbon 2 as its anomeric carbon atom.

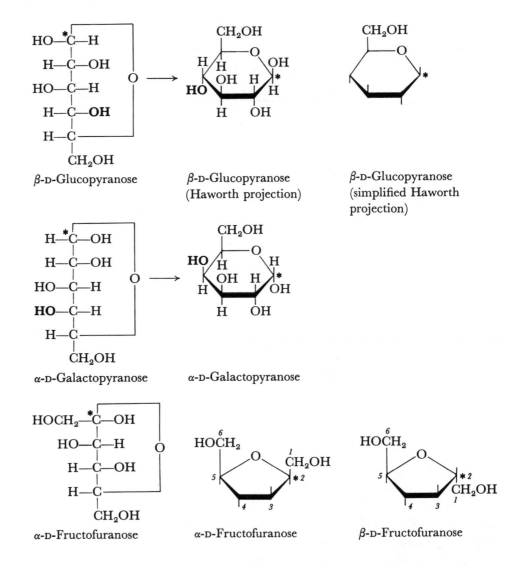

β-D-Glucopyranose

β-D-Glucopyranose
(Haworth projection)

β-D-Glucopyranose
(simplified Haworth
projection)

α-D-Galactopyranose

α-D-Galactopyranose

α-D-Fructofuranose

α-D-Fructofuranose

β-D-Fructofuranose

The actual conformation of the pyranose and furanose rings is not planar, as the Haworth projection makes it appear. Instead, the pyranose ring exists in "chair" and "boat" forms as shown below:

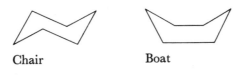

Chair

Boat

Problems 1–7

Which of the structures below represent carbohydrates? Classify each carbohydrate in terms of the number of carbon atoms and decide whether it is an aldose or ketose, pyranose or furanose, and α or β isomer.

1.

CH₂OH

2.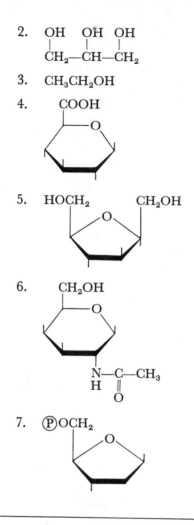

     OH   OH   OH

     | | |

    CH_2—CH—CH_2

3. CH_3CH_2OH

4.   COOH

5. $HOCH_2$ CH_2OH

6. CH_2OH

               N—C—CH_3

               H ‖

                  O

7.   (P)OCH_2

REDUCING SUGARS

A sugar can reduce Cu^{+2} if its anomeric carbon has a free hydroxyl group; such sugars are termed *reducing sugars*. The monosaccharides glucose, galactose, fructose, and ribose are all reducing sugars.

Many of the urine dipsticks measure reducing sugars in general and are not specific for glucose. A positive urine reducing-sugar test must be pursued further to identify the particular compound, whether glucose, fructose, galactose, or even ascorbate, among many other compounds.

DISACCHARIDES

Disaccharides consist of two monosaccharides joined by a glycosidic bond. Each glycosidic bond is classified as α or β and is numbered according to the positions of the carbon atoms it links.

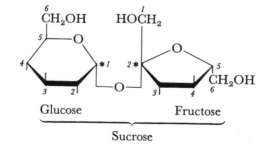

Glucose   Fructose

Sucrose

In sucrose, the α-anomeric carbon 1 of glucose joins the β-anomeric carbon 2 of fructose. Because neither anomeric carbon has a free hydroxyl group, sucrose will not reduce Cu^{+2} to Cu^+, nor can it mutarotate. Sucrose, or table sugar, comes primarily from sugar cane and sugar beets.

Maltose, a dimer of glucose linked by an α-1,4 glycosidic bond, is produced during gastrointestinal starch digestion.

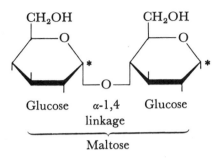

Lactose, found naturally only in milk products, consists of β-galactose with a β-1,4 linkage to β-glucose.

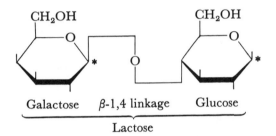

The intestinal mucosa contains the enzymes sucrase, maltase, and lactase, which cleave the glycosidic bonds of sucrose, maltose, and lactose, respectively. Sucrase, for example, cleaves sucrose to glucose and fructose. Lactase deficiency, the most common intestinal disaccharidase deficiency, causes osmotic diarrhea (due to the presence of the undigested lactose in the intestine) and excessive gas production as a result of lactose fermentation by intestinal bacteria.

POLYSACCHARIDES

Polysaccharides are carbohydrate polymers with more than ten monosaccharide units. *Oligosaccharides* contain two to ten monosaccharide residues.

Cellulose, the structural polysaccharide of plants, is a long, unbranched chain of glucose units with β-1,4 linkages. Only certain bacteria possess the

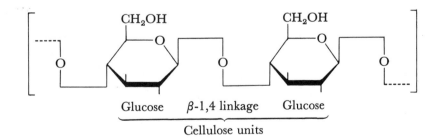

Cellulose units

Figure 7-1

Comparison of amylopectin and glycogen. Amylopectin has branches every 24 to 30 glucose units, whereas glycogen has branches every 8 to 12 glucose residues.

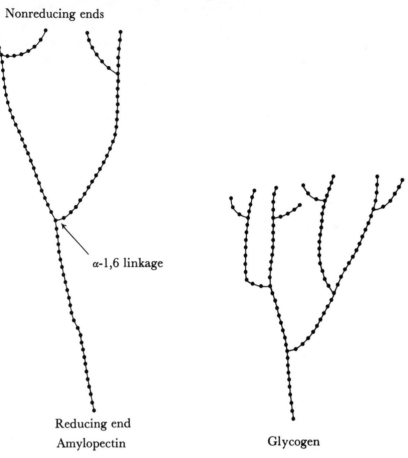

Nonreducing ends

α-1,6 linkage

Reducing end

Amylopectin

Glycogen

Figure 7-2

Glycogen phosphorylase and glucan transferase convert glycogen into a limit dextrin.

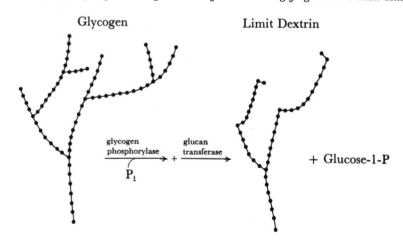

Glycogen

Limit Dextrin

glycogen
phosphorylase

P_1

glucan
transferase

+ Glucose-1-P

78 7. Structure and Properties of Carbohydrates

cellulase enzyme that is required to cleave β-1,4 glycosidic bonds. Cows depend on bacteria in their rumen to digest cellulose.

Starch, the energy-storing polysaccharide of plants, is a mixture of amylose and amylopectin. Amylose is a long, unbranched glucose polymer with α-1,4 bonds. Amylopectin, on the other hand, has α-1,6 branch linkages spaced every 24 to 30 glucose residues in its α-1,4 chain (Fig. 7-1). One end of the chain has a free OH group on the anomeric carbon, making it the reducing end of the polymer. The other ends of amylopectin are nonreducing, because their anomeric carbons are linked in glycosidic bonds.

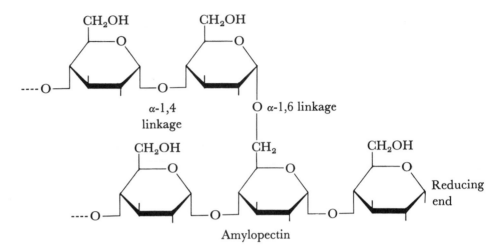

Amylopectin

Glycogen, the glucose-storing polysaccharide of animals, resembles amylopectin in having α-1,6 branches from an α-1,4 chain. Glycogen, however, is more highly branched, with α-1,6 linkages every 8 to 12 glucose residues, as shown in Figure 7-1.

α-Amylase, produced by the salivary glands and pancreas, randomly hydrolyzes the α-1,4 linkages of dietary polysaccharides. α-Amylase hydrolyzes amylose to a mixture of maltose and glucose. When α-amylase digests dietary amylopectin and glycogen, it cannot attack the α-1,6 linkages. Thus, after cleaving the branch chains down to the α-1,6 bonds, it leaves a glucose polymer skeleton called a *limit dextrin*.

Amylo-1,6-glucosidases, the debranching enzymes, hydrolyze these α-1,6 bonds and allow α-amylase to continue its task on the internal α-1,4 linkages.

Glycogen phosphorylase in human tissues removes glucose from tissue glycogen by phosphorylation to glucose-1-phosphate. Glucan transferase then transfers the final several glucose residues of the branch to another arm of the polymer, leaving a limit dextrin as shown in Figure 7-2. Amylo-1,6-glucosidase must then debranch this limit dextrin to allow glycogen phosphorylase to degrade the remainder of the glycogen molecule.

MUCOPOLY-SACCHARIDES

Unlike starch and glycogen, the *mucopolysaccharide* polymers are built up from more than one type of carbohydrate unit. Their components may include aminated, sulfated, and *N*-acetylated sugars.

Mucopolysaccharides abound in the ground substance of connective tissue, where they play a structural role. The loss of mucopolysaccharides from the intervertebral discs predisposes to degenerative disc disease, which is extremely common in the United States.

The acidic mucopolysaccharides contain sugar acids, such as glucuronic acid.

Hyaluronic acid is an acidic mucopolysaccaride made up of the repeating disaccharide unit of glucuronic acid joined to *N*-acetylglucosamine. The staphylococci, for example, can readily invade connective tissue, because they secrete hyaluronidase, an enzyme that cleaves hyaluronic acid.

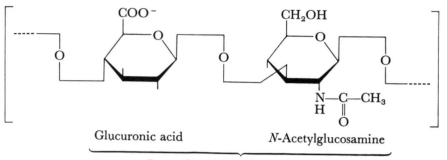

Glucuronic acid　　　　　*N*-Acetylglucosamine

Repeating unit of hyaluronic acid

Heparin, a sulfated, acidic mucopolysaccaride, is normally present in most cells and is commonly administered as an anticoagulant. It consists of an *O*-sulfated glucuronic acid bound to an *N*-sulfated glucosamine that has a second *O*-sulfate group.

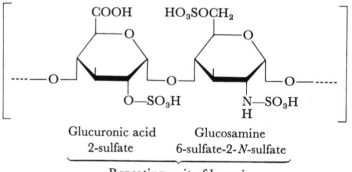

Glucuronic acid　　　　Glucosamine
2-sulfate　　　　6-sulfate-2-*N*-sulfate

Repeating unit of heparin

Chondroitin sulfate A contains the repeating disaccharide unit of glucuronic acid joined to *N*-acetylgalactosamine-4-sulfate.

The mucopolysaccharidoses are a class of rare, inherited disorders of mucopolysaccharide metabolism. They cause bony deformities and abnormalities of the facial features as a result of mucopolysaccharide deposits in connective tissue. A common screening test for the mucopolysaccharidoses is to assay the urine for mucopolysaccharides, which are secreted in excess in these disorders.

Problem 8

Which of the polymers built upon the repeating disaccharide units shown below would appear in the urine of patients with one of the mucopolysaccharidoses? Name those sugar units that do not form mucopolysaccharides.

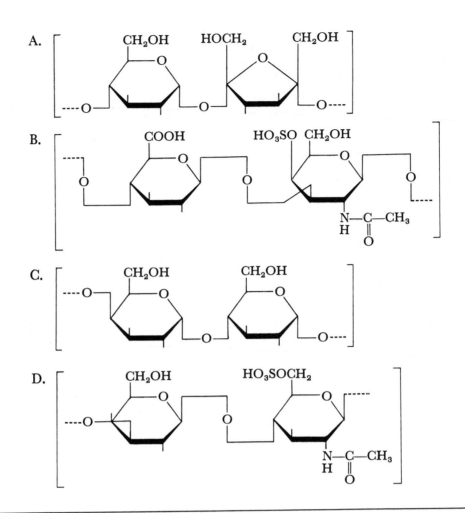

Problem 9

Tests show that your new patient has marked glycosuria (excessive reducing sugars in the urine). Which of the following situations could be associated with a positive test for excessive amounts of reducing sugars in the urine?

A. Galactosemia in which the excess serum galactose spills into the urine
B. Dietary pentosuria resulting from an overdose of Hawaiian punch, which contains an aldopentose
C. The lab tech added sucrose to the urine sample rather than to his coffee
D. Diabetic ketoacidosis with excess urinary glucose
E. Essential fructosuria manifested by excess urinary fructose

---

Problems 10–12

Choose the foods below that must be eliminated from the diets of persons with the metabolic diseases listed in Problems 10–12:

A. Milk products
B. Honey, fruits
C. Table sugar

10. Hereditary fructose intolerance (avoid fructose).
11. Sucrase deficiency (avoid sucrose).
12. Lactase deficiency (avoid lactose).

α-Amylase hydrolysis of amylopectin results in:

A. Complete cleavage to glucose and maltose
B. Limit dextrin formation
C. Removal of α-1,6 branch linkages
D. Complete cleavage to glucose and maltose in the presence of α-amylo-1,6-glucosidase

Describe the structures below and the type of reaction (e.g., phosphorylation, reduction, or whatever):

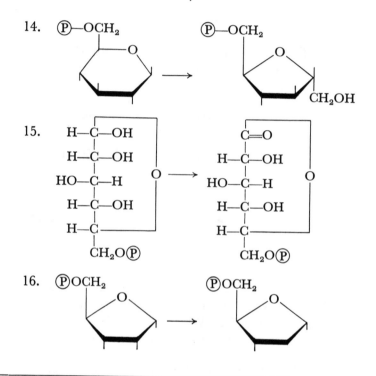

14.

15.

16.

ANSWERS

1. Aldohexose in pyranose form. It is the α-anomer, because the OH on the anomeric carbon, carbon 1, is directed downward. This is mannose, a sugar present in humans.

2. Triose sugar alcohol, glycerol. It does not form a ring structure nor does it have anomers.

3. This is ethanol, which is not a carbohydrate.

4. Sugar acid of an aldohexopyranose, glucuronic acid. It is the β-anomer, because the OH on the anomeric carbon is pointing upward.

5. Ketohexose in furanose form. This is the α-anomer, because the OH on carbon 2, the anomeric carbon, points downward. This is fructose.

6. Aldohexose in pyranose form, β-anomer. Replacing the OH on carbon 2 is an N-acetyl group. This is N-acetyl-β-D-galactosamine.

7. Aldopentose in furanose form, β-anomer. Phosphate replaces the OH on carbon 5, and carbon 2 lacks a hydroxyl group. This is β-D-2-deoxyribose-5-phosphate.

8. A. No. The left-hand residue is glucose and the right-hand residue is fructose. Thus, the disaccharide unit is sucrose.

B. Yes. On the left is the sugar acid, glucuronic acid. On the right is *N*-acetylglucosamine-4-sulfate. This compound is chondroitin sulfate A.

C. No. On the left is galactose while on the right is glucose. Hence, this disaccharide unit is lactose (it is easy to remember that *lactose* contains *galactose*).

D. Yes. On the left is galactose, which is β-1,4 linked to *N*-acetylglucosamine-6-sulfate. This compound happens to be keratan sulfate I. It contains no sugar acids and thus is not an acidic mucopolysaccharide.

9. A, B, D, and E. Sucrose is a nonreducing sugar because it lacks a free hydroxyl group on either anomeric carbon; it will therefore not give a positive test for glycosuria.

10. B, C.

11. C.

12. A.

13. B, D.

14. β-Glucopyranose-6-phosphate $\longrightarrow$ β-fructofuranose-6-phosphate. Isomerization.

15. On the left is glucose-6-phosphate. On the right, the OH group of carbon 1 has been oxidized to C=O, creating a lactone (you are not expected to name this compound, which is gluconolactone-6-phosphate).

16. On the left is α-ribose-5-phosphate. On the right is 2-deoxy-α-ribose-5-phosphate. This is a reduction reaction.

REFERENCES

Barker, R. *Organic Chemistry of Biological Compounds.* Englewood Cliffs, N.J.: Prentice-Hall, 1971. Pp. 139–157, 196–203.

Bhagavan, N. V. *Biochemistry—A Comprehensive Review.* Philadelphia: Lippincott, 1974. Pp. 131–153.

Lehninger, A. L. *Biochemistry: The Molecular Basis of Cell Structure and Function* (2nd ed.). New York: Worth, 1975. Pp. 249–276.

White, A., Handler, P., and Smith, E. L. *Principles of Biochemistry* (5th ed.). New York: McGraw-Hill, 1973. Pp. 13–36, 42–57, 412–416.

8 Structure and Properties of Lipids

Lipids are compounds that are soluble in nonpolar solvents such as ether and benzene. Although certain lipids contain ionized groups (e.g., phosphate or choline), the bulk of any lipid molecule is nonpolar. Amino acids, proteins, carbohydrates and nucleotides are too highly polar and ionized to be soluble in these "lipid" solvents.

Structurally, the lipids are quite diverse; there is no common subunit in their structure. The primary building blocks in human lipids are fatty acids, glycerol, sphingosine, and sterols.

FATTY ACIDS

All fatty acids have a single carboxyl group at the end of a hydrocarbon chain, which makes them weak acids. Acetic acid, CH_3COO^-, is the simplest fatty acid. The three-carbon fatty acid (C3) is propionic acid, $CH_3CH_2COO^-$.

Most natural fatty acids have an even number of carbon atoms.

The hydrocarbon chain of fatty acids, represented by $RCOO^-$, can be either saturated (i.e., lacking carbon-carbon double bonds) or unsaturated.

The two most abundant saturated fatty acids in humans are palmitic acid (C16) and stearic acid (C18).

Oleic acid (C18) and palmitoleic acid (C16) compose the bulk of the monounsaturated, or monoenoic, fatty acids in humans. Both have a carbon-carbon double bond between carbons 9 and 10.

Polyunsaturated, or polyenoic, fatty acids include linoleic acid (C18) with two double bonds, linolenic acid (C18) with three, and arachidonic acid (C20) with four double bonds.

Polyunsaturated fatty acids are the precursors for the prostaglandins. As shown below, part of the straight chain of arachidonic acid is folded into a five-carbon ring while two of its double bonds are oxygenated to generate prostaglandin E_2 (PGE_2). The bonds drawn as dots project downward perpendicular to the page.

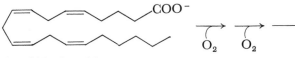

Arachidonic acid

PGE_2

85

ACYLGLYCEROLS

The *acylglycerols* are esters of fatty acids bound to the sugar alcohol glycerol. They are also called *neutral fats*, because the carboxyl groups of the fatty acids are bound in ester linkage and can no longer function as acids.

The triacylglycerols, or *triglycerides*, are the principal storage fats in humans. They are named according to their fatty-acid content; tripalmitin contains glycerol and three palmitate chains, while stearodiolein contains glycerol with one stearate and two oleate chains.

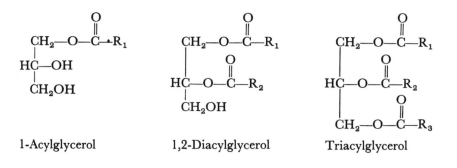

1-Acylglycerol   1,2-Diacylglycerol   Triacylglycerol

In humans, triglycerides are hydrolyzed to glycerol and free fatty acids (FFA) by lipase enzymes.

In industry, neutral fats are hydrolyzed with NaOH, or *saponified*, to create water-soluble fatty-acid soaps ($RCOO^-Na^+$) and glycerol. Soaps have a detergent action because after the nonpolar R groups in the soaps bind to lipids in the skin or clothing, the ionized carboxyl group can pull this bound lipid into the water phase.

Calcium binds to free fatty acids in the intestine to generate insoluble calcium-fatty-acid soaps, thereby blocking their absorption. In acute pancreatitis, lipase released from the pancreas into the bloodstream hydrolyzes triacylglycerols and creates calcium-fatty-acid soaps.

PHOSPHO-GLYCERIDES

The *phosphoglycerides* are phosphate esters of diglycerides. Glycerol-3-phosphate is the structural backbone of the phosphoglycerides. Two fatty acids are esterified to glycerol-3-phosphate to produce the *phosphatidic acids*, which are intermediates in the synthesis of triacylglycerols and various other phosphoglycerides.

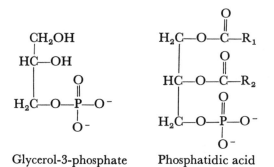

Glycerol-3-phosphate   Phosphatidic acid

By esterifying choline, or trimethylethanolamine $HOCH_2CH_2{}^+N(CH_3)_3$, to the phosphoric acid portion of phosphatidic acid, one gets phosphatidylcholine. Also called lecithin, the phosphatidylcholines play an essential role in reducing surface tension in lung alveoli; they are surfactants or surface-acting agents. Respiratory distress syndrome (RDS) of the newborn, which is common in premature infants, results from a lack of this surfactant in the lung. The lung is stiff, expands with difficulty, and has many collapsed portions. To predict the likelihood of RDS in high-risk pregnancies, obstetricians commonly perform amniocentesis for laboratory determination of the ratio of phosphatidylcholine (lecithin) to sphingomyelin in the amniotic fluid (L/S ratio). The higher the L/S ratio, the more surfactant is present to allow the lung to expand normally.

Other important human phosphoglycerides are phosphatidylserine and phosphatidylethanolamine. The common serologic test for syphilis, the Venereal Disease Research Laboratory (VDRL) test, utilizes cardiolipin, a diphosphatidyl glycerol, as the antigen.

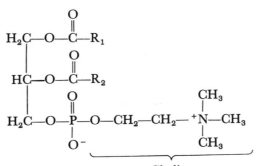

Phosphatidylcholine

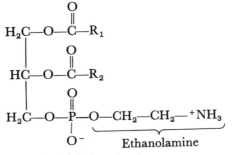

Phosphatidylethanolamine

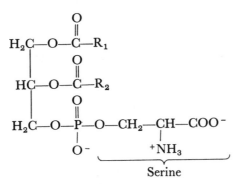

Phosphatidylserine

The highly polar phosphate, choline, and serine groups of the phosphoglycerides make these compounds water-soluble, while their fatty acyl groups confer solubility in nonpolar agents. Hence, they can serve to cement lipids in membranes and lipoproteins to the polar proteins and carbohydrates.

SPHINGOLIPIDS

The *sphingolipids* are so named because they all contain sphingosine or one of its derivatives. The structure of sphingosine may be easily identified, because it contains a long-chain, monounsaturated alcohol bound to ethanolamine:

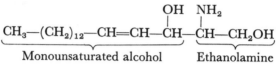

In addition to sphingosine, all sphingolipids contain a fatty acid. *None contains glycerol.*

Sphingolipids abound in the nervous system as components of myelin and other structural lipids. They occur to a lesser extent in the liver, spleen, and bone marrow.

Ceramides, the simplest sphingolipids, consist of a fatty acid bound to sphingosine. In humans, ceramides function principally as intermediates in the synthesis of other sphingolipids; all other sphingolipids contain ceramide.

$$CH_3-(CH_2)_{12}-CH=CH-\underset{\underset{\displaystyle \text{Ceramide}}{|}}{CH}-\underset{\underset{\displaystyle \overset{|}{N}-\overset{\displaystyle \overset{O}{\|}}{C}-R}{|}}{CH}-CH_2OH$$

By joining choline phosphate or ethanolamine phosphate to ceramides, one generates the *sphingomyelins*, which are important components of the myelin sheath surrounding the fastest conducting nerve fibers.

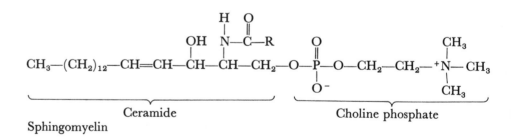

Cerebrosides consist of a hexose sugar, such as glucose or galactose, bound to a ceramide. These ceramide-monosaccharides are also part of the myelin sheath.

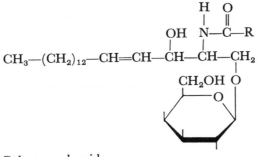

Galactocerebroside

Gangliosides differ from cerebrosides in that they contain additional sugar residues, such as glucose or galactose, and amino sugars, such as *N*-acetylgalactosamine and *N*-acetylneuraminic acid.

Sulfatides are sulfated cerebrosides, or cerebroside-sulfate esters.

Problems 1–6

Describe the lipids below in terms of these components: saturated and unsaturated fatty acids, glycerol, phosphate, sphingosine, carbohydrate, choline, ethanolamine, and serine. Then classify each as either a fatty acid, prostaglandin, acylglycerol, phosphoglyceride, ceramide, cerebroside, sphingomyelin, or ganglioside.

1.

$$H_2C-O-\overset{\overset{\displaystyle O}{\|}}{C}-(CH_2)_{16}-CH_3$$
$$H-\overset{\displaystyle |}{C}-OH$$
$$CH_2OH$$

2.

$$CH_3-(CH_2)_{12}-CH=CH-\overset{\overset{\displaystyle OH}{|}}{C}H-\overset{\overset{\displaystyle H}{|}}{\underset{}{C}}H-CH_2OH$$

with $\overset{H}{N}-\overset{\overset{\displaystyle O}{\|}}{C}-(CH_2)_{14}-CH_3$

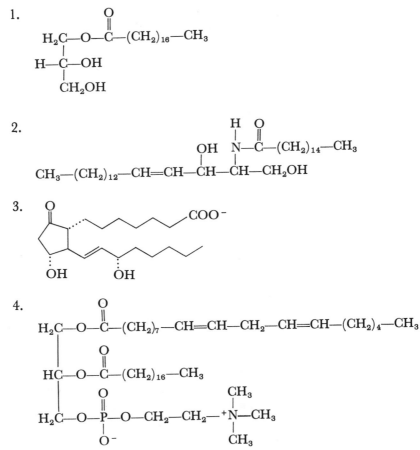

3.

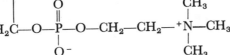

$$COO^-$$

OH   OH

4.

$$H_2C-O-\overset{\overset{\displaystyle O}{\|}}{C}-(CH_2)_7-CH=CH-CH_2-CH=CH-(CH_2)_4-CH_3$$
$$HC-O-\overset{\overset{\displaystyle O}{\|}}{C}-(CH_2)_{16}-CH_3$$
$$H_2C-O-\overset{\overset{\displaystyle O}{\|}}{\underset{\underset{\displaystyle O^-}{|}}{P}}-O-CH_2-CH_2-{}^+\overset{\overset{\displaystyle CH_3}{|}}{\underset{\underset{\displaystyle CH_3}{|}}{N}}-CH_3$$

89

5.
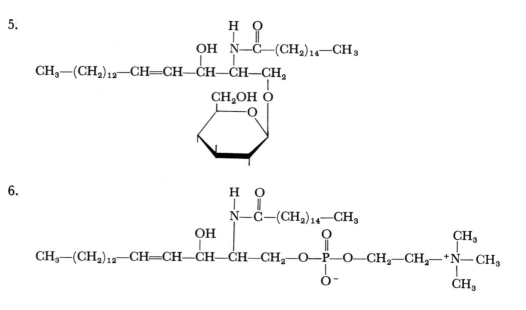

$$CH_3-(CH_2)_{12}-CH=CH-\underset{\underset{CH_2OH}{|}}{\overset{\overset{OH}{|}}{CH}}-\underset{\underset{\underset{O}{|}}{\overset{\overset{H}{|}}{N}}}{CH}-\underset{\overset{H\;O}{\overset{|\;\;\|}{N-C}}}{CH_2}-(CH_2)_{14}-CH_3$$

6.

$$CH_3-(CH_2)_{12}-CH=CH-\overset{\overset{OH}{|}}{CH}-\underset{\overset{\overset{H\;O}{|\;\;\|}}{N-C-(CH_2)_{14}-CH_3}}{CH}-CH_2-O-\overset{\overset{O}{\|}}{\underset{\underset{O^-}{|}}{P}}-O-CH_2-CH_2-\overset{\overset{CH_3}{|}}{\underset{\underset{CH_3}{|}}{{}^+N}}-CH_3$$

TERPENES

The *terpenes* are a class of isoprene polymers. Examples of terpene derivatives in humans include the cholesterol precursors (squalene, geraniol, and farnesol), provitamin A (β-carotene), and vitamin A_1 (retinol).

STEROIDS

The broad category of *steroids* includes the steroid hormones, sterols, and bile acids.

The essential structural nucleus of the steroids consists of three fused cyclohexane rings joined to a cyclopentane ring. Except for the estrogens, steroids do not contain aromatic rings. Carbons 3 and 17 always have side groups.

Steroid nucleus

Estrogens, the ovarian steroids, contain 18 carbon atoms (carbon 18 is found in a methyl group). Unlike other steroids, one ring of the steroid nucleus of estrogen is aromatic. Estradiol (shown below) has OH groups attached to carbons 3 and 17.

Androgens, produced in the adrenal cortex and the testes, have 19 carbon atoms (carbons 18 and 19 are in methyl groups). Testosterone is one of the more potent androgens. Because of its keto group at carbon 3, testosterone has a double bond in its initial cyclohexane ring. Dehydroepiandrosterone (DEA), unlike testosterone, is a 17-ketosteroid. Adrenal hyperplasia will increase the urinary excretion of 17-ketosteroids.

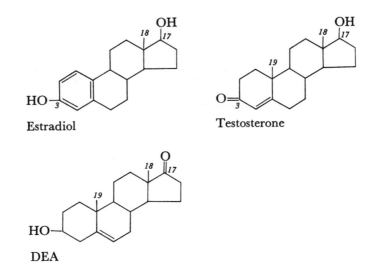

Estradiol

Testosterone

DEA

Progesterone, synthesized in the corpus luteum, has 21 carbons, as do the adrenocortical steroids such as corticosterone and cortisol. Progesterone has an acetyl group joined at carbon 17.

The adrenal cortex produces *glucocorticoids*, which raise the serum glucose level, and *mineralocorticoids*, which promote renal sodium retention. Cortisol is a potent glucocorticoid with weak mineralocorticoid activity, whereas aldosterone is a potent mineralocorticoid (presumably due to its aldehyde group at carbon 18) but a weak glucocorticoid. A hydroxyl or keto group at carbon 11 is found to correlate with glucocorticoid activity. The urinary 17-hydroxycorticosteroid assay detects all 21-carbon steroids with a 17-OH group, such as cortisol.

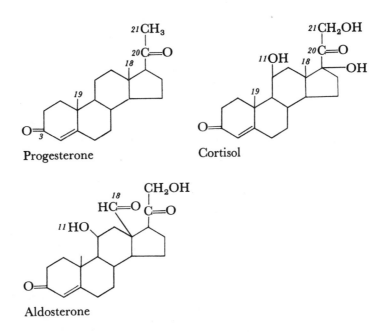

Progesterone

Cortisol

Aldosterone

The *bile acids* are 24-carbon steroids secreted into the bile to emulsify dietary fats. They have a five-carbon side chain at position 17 that contains a carboxyl group, making them acidic. Cholic acid is a major human bile acid.

The *sterols* are steroids with 27 to 29 carbon atoms and an OH group at carbon 3. Cholesterol, the major human sterol, is the precursor to all the steroid hormones. In addition, it occurs in high concentration in the brain. Most cholesterol in blood is bound to unsaturated fatty acid through the OH group at carbon 3 to form cholesterol esters. The vitamin D precursors, ergosterol and 7-dehydrocholesterol, are also sterols.

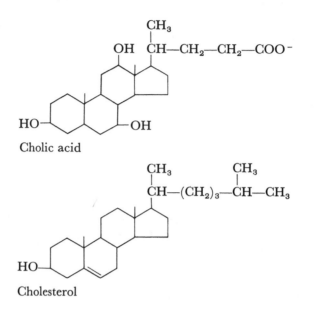

Cholic acid

Cholesterol

LIPID DIGESTION

Lipid digestion begins in the duodenum, where bile salts and pancreatic lipase and phospholipases first meet the partially digested food. Lipase hydrolyzes triacylglycerols into free fatty acids, glycerol, monoacylglycerols, and some diacylglycerols. Phospholipases remove fatty acids from phosphoglycerides. Bile salts solubilize or emulsify this mixture of free fatty acids, monoacylglycerols, and diacylglycerols into droplets less than 1 micron in diameter termed *micelles,* which are readily absorbed throughout the small intestine.

Inside the intestinal mucosal cells, triacylglycerols are re-synthesized and combined with β-lipoproteins, phosphoglycerides, and cholesterol to form *chylomicrons,* which enter the lymphatics and travel through the thoracic duct to reach the bloodstream.

Unlike long-chain fatty acids, the medium- and short-chain fatty acids (i.e., those with less than 12 carbons) can be well absorbed without bile salts. They enter the portal venous system directly and travel to the liver, instead of traveling through the lymphatics as triacylglycerols in chylomicrons. Medium- and short-chain fatty-acid preparations are used as a source of fatty acids for patients with lipid malabsorption disorders, such as cystic fibrosis.

Bile salts also emulsify cholesterol to hasten its absorption. Once absorbed, cholesterol is esterified to unsaturated fatty acids to create cholesterol esters.

Cholestyramine is a resin used clinically to lower the serum cholesterol level. It binds to bile salts, thereby blocking cholesterol absorption and augmenting the hepatic transformation of cholesterol to bile salts.

LIPOPROTEINS

Lipids must bind to proteins to make them water-soluble for transport in the blood. Free fatty acids, for example, avidly bind to serum albumin and will displace albumin-bound drugs from their binding sites.

Two laboratory techniques are used to separate lipoproteins from one another: ultracentrifugation separates them according to their differing densities and electrophoresis separates them on a basis of their varying net charges.

Chylomicrons are the least dense lipoproteins, because they consist mainly of triglycerides with small amounts of cholesterol, phospholipids, and proteins. They do not migrate when subjected to electrophoresis, because of their high triglyceride content (triacylglycerols have no charge). After a fatty meal, the blood appears milky due to the high concentration of chylomicrons. Lipoprotein lipase hydrolyzes triglycerides inside the chylomicrons, and it removes chylomicrons from the circulation. Heparin, an anticoagulant, also helps to clear chylomicrons from the blood, perhaps by stimulating lipoprotein lipase. The inherited absence of lipoprotein lipase causes hyperchylomicronemia, termed Fredrickson's type-I hyperlipoproteinemia.

Very-low-density lipoproteins (VLDL) also contain principally triglycerides, but they have a greater protein, phospholipid, and cholesterol content than chylomicrons. Their protein and phospholipid content makes them charged so that they migrate just before the β-globulins in electrophoresis; hence, they are termed *pre-β-lipoproteins*. VLDL are synthesized in the liver. This VLDL fraction is markedly elevated in type-IV hyperlipoproteinemia.

Low-density lipoproteins (LDL) contain mainly cholesterol, in contrast to the content of chylomicrons and VLDL, which is mainly triglyceride. LDL also contain appreciable amounts of proteins, phospholipids, and triglycerides. Because they migrate with the β-globulins, they are termed *β-lipoproteins*. The LDL fraction is markedly elevated in Type-II hyperlipoproteinemia.

High-density lipoproteins (HDL) contain mainly protein and phospholipid. They contain significant amounts of cholesterol, but they have little triglyceride. The high-protein, low-triglyceride content makes them very dense. They are termed *α-lipoproteins* and are separated from other lipoproteins by electrophoresis.

Problem 7

Choose the *incorrect* phrase describing the estrogens:

A. Eighteen carbon atoms
B. The only steroid class that has an aromatic ring
C. Single methyl group attached to steroid nucleus
D. OH at carbon 11

Problem 8

Cholesterol is:

A. an acidic sterol
B. a 17-ketosteroid
C. a precursor for all the steroid hormones
D. a 17-OH corticosteroid

Match the lipoprotein density categories below to the descriptions given in Problems 9–14:

A. chylomicrons   C. LDL
B. VLDL   D. HDL

9. Composed mainly of protein and phosphoglyceride
10. Cleared from plasma by lipoprotein lipase
11. Consist mainly of triglyceride and are synthesized in the liver
12. Consist mainly of cholesterol
13. β-Lipoproteins
14. Pre-β-lipoproteins

Match the structures below to their descriptions given in Problems 15–18. More than one choice may be used for each question.

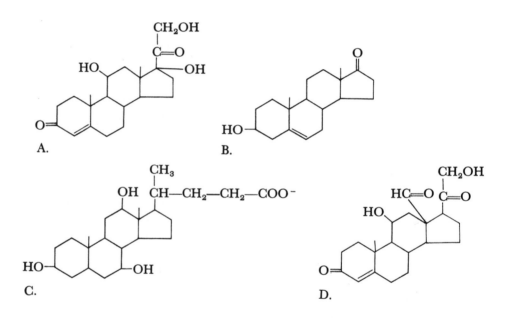

15. 17-Ketosteroid(s). Elevated 24-hour urinary excretion suggests adrenal hyperplasia.
16. Bile acid(s). Deficiency impairs micelle formation.
17. 21-Carbon corticosteroid(s). Aldehyde at carbon 18 confers potent mineralocorticoid activity.
18. 17-OH corticosteroid(s). 11-Hydroxy group confers glucocorticoid activity.

ANSWERS

1. Saturated fatty acid (C18) esterified to glycerol. Monoacylglycerol or monoglyceride.
2. Sphingosine joined to a 16-carbon, saturated fatty acid (palmitic acid) by N-acyl linkage. (Sphingosine can be recognized by identifying its monounsaturated alcohol bound to ethanolamine.) A sphingolipid containing only sphingosine and a fatty acid is a ceramide.

3. Oxygenated derivative of a 20-carbon, polyenoic fatty acid. Its five-membered ring and two aliphatic chains should identify this as a prostaglandin (PGE_1).

4. Eighteen-carbon, doubly unsaturated fatty acid (linoleic acid) plus an 18-carbon, saturated fatty acid (stearic acid) esterified to glycerol phosphorylcholine (phosphatidylcholine).

5. Sixteen-carbon, saturated fatty acid (palmitic acid) joined by N-acyl linkage to sphingosine, which in turn is bound by glycosidic linkage to glucose. This is a glucocerebroside.

6. Palmitic acid and phosphorylcholine bound to sphingosine. This is a sphingomyelin.

7. D. (Only the steroids with glucocorticoid activity have an OH at carbon 11.)

8. C.

9. D.

10. A.

11. B. (Chylomicrons also consist mainly of triglyceride, but they are synthesized by the intestine.)

12. C.

13. C.

14. B.

15. B. (Dehydroepiandrosterone.)

16. C. (Note the carboxylic acid group; cholic acid.)

17. D. (Aldosterone.)

18. A. (Cortisol.)

REFERENCES

Bhagavan, N. V. *Biochemistry—A Comprehensive Review*. Philadelphia: Lippincott, 1974. Pp. 592–635, 662–675.

Fredrickson, D., and Levy, R. Familial Hyperlipoproteinemia. In Stanbury, J., Wyngaarden, J., and Fredrickson, D. (Eds.), *The Metabolic Basis of Inherited Disease* (3rd ed.). New York: McGraw-Hill, 1972.

Lehninger, A. L. *Biochemistry: The Molecular Basis of Cell Structure and Function* (2nd ed.). New York: Worth, 1975. Pp. 279–300.

White, A., Handler, P., and Smith, E. L. *Principles of Biochemistry* (5th ed.). New York: McGraw-Hill, 1973. Pp. 59–85, 542–546, 559–585.

9 Structure and Properties of Nucleic Acids

The fundamental components of nucleic acids are the pyrimidine and purine bases, the pentose sugars ribose and 2-deoxyribose, and phosphoric acid.

PYRIMIDINES

The *pyrimidine bases* contain the six-membered ring with two nitrogen atoms that constitutes the compound, pyrimidine. The three major pyrimidines found in man—cytosine, uracil, and thymine—have oxygenated pyrimidine rings. Thymine, not to be confused with thiamine or vitamin B_1, has a methyl group at position 5. Cytosine has an amino group at position 4.

Oxygenated pyrimidines and purines exist in two different tautomeric forms, differing only with respect to the location of a proton. *Tautomers* are isomers that are freely interconvertible and exist in dynamic equilibrium under normal conditions. The keto or lactam (NHC=O) tautomer of cytosine is generally more prevalent than the enol or lactim (N=COH) tautomer at neutral pH.

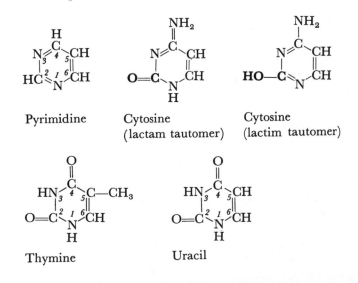

Pyrimidine   Cytosine (lactam tautomer)   Cytosine (lactim tautomer)

Thymine   Uracil

PURINES

The *purine bases* consist of a pyrimidine ring fused to an imidazole ring (histidine also has an imidazole ring). The two major purines in human nucleic acids are adenine and guanine. Adenine lacks a hydroxyl or a keto group and therefore cannot exhibit keto-enol tautomerism; it has an amino group at position 6. Guanine has a keto group at position 6 and an amino group at position 2.

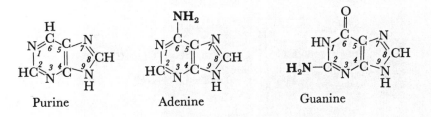

Purine   Adenine   Guanine

Inosine, hypoxanthine, xanthine, and uric acid are purines created during the degradation of adenine and guanine.

The stimulant drugs caffeine (in coffee and tea) and theobromine (in chocolate) are methylxanthine compounds.

Problems 1–7

Which of the structures below are pyrimidines or purines? For those that are purines or pyrimidines, decide whether the keto or enol (lactam or lactim) tautomer is shown and identify the groups attached to the ring.

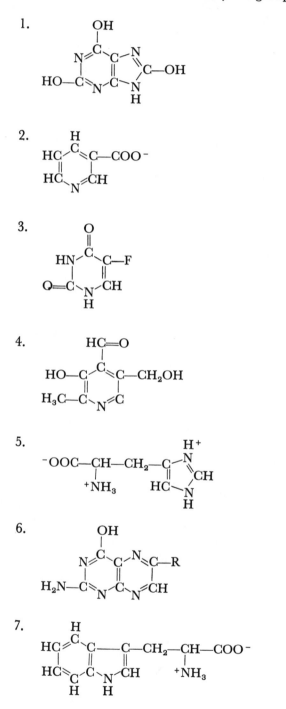

NUCLEOSIDES

A *nucleoside* consists of a purine or pyrimidine base bound to a pentose sugar. The nucleosides of ribose with adenine, guanine, cytosine, thymine, and uracil are called adenosine, guanosine, cytidine, thymidine, and uridine, respectively. If deoxyribose is present rather than ribose, the prefix *deoxy* is used, as in deoxyuridine.

In numbering the atoms in nucleosides, a superscript prime is used to denote the atoms in the pentose. Thus, in cytidine, atom 1 of cytosine is joined by a glycosidic bond to carbon 1' of ribose.

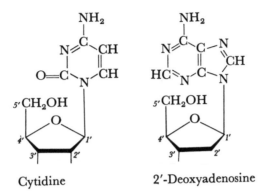

Cytidine   2'-Deoxyadenosine

NUCLEOTIDES

The terms "nucleotide" and "nucleoside phosphate" can be used interchangeably. In a *nucleotide*, a phosphate group is bound to the pentose sugar of a nucleoside. This phosphate group makes nucleotides strongly acidic.

The 5'-nucleoside monophosphate (NMP) of adenosine is called adenylic acid or adenosine monophosphate (AMP). Guanosine monophosphate (GMP) is called guanylic acid, cytidylic acid is CMP, uridylic acid is UMP, and thymidylic acid is TMP.

The 5'-nucleoside diphosphates (NDPs) are ADP, GDP, CDP, UDP, and TDP.

The 5'-nucleoside triphosphates (NTPs) are ATP, GTP, CTP, UTP, and TTP.

In 3',5'-cyclic AMP, the phosphate group joins to both the 3' and 5' carbons of ribose. This compound acts as a "second messenger" for many of the polypeptide hormones. For example, glucagon, a polypeptide hormone and first messenger, stimulates adenyl cyclase in liver cell membranes to cleave ATP to 3',5'-cyclic AMP. This cyclic nucleotide in turn acts as a second messenger

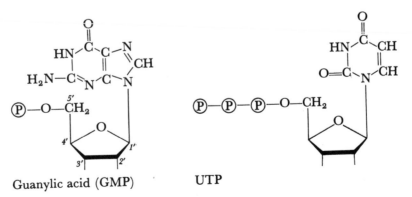

Guanylic acid (GMP)   UTP

99

to activate glycogen phosphorylase and inhibit glycogen synthase, thereby freeing glucose so that its level in the blood may be maintained.

Problems 8–10

Match the structures below to their descriptions given in Problems 8–10:

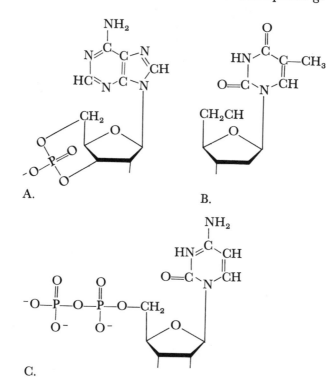

A.

B.

C.

8. A cytidine nucleotide
9. 3′,5′-Cyclic AMP
10. Thymidine

POLYNUCLEOTIDES AND THE NUCLEIC ACIDS

The nucleotides of a polynucleotide chain are linked to one another in 3′,5′-phosphodiester bonds; phosphoric acid forms a phosphate ester to connect the 3′ hydroxyl group of one pentose to the 5′ carbon on another pentose.

In the schematic structure for polynucleotides (see facing page), P represents phosphoric acid and A, C, G, U, and T represent the nucleosides. In naming the sequence of a polynucleotide, the 5′ end is written to the left of the 3′ end; pU designates a 5′-phosphate, whereas Up is a 3′-phosphate. Thus, the above dinucleotide is pG-Cp. If it contained deoxyribose instead of ribose, it would be written dpG-Cp.

Ribonucleic acid (RNA) is a single-stranded polynucleotide whose pentose sugar is ribose. *Deoxyribonucleic acid (DNA)* is a double-stranded polynucleotide with deoxyribose as its pentose.

The base content of DNA displays three sets of equivalent pairs:

1. A + G = T + C (purine content = pyrimidine content)
2. A = T
3. G = C

Schematic structure

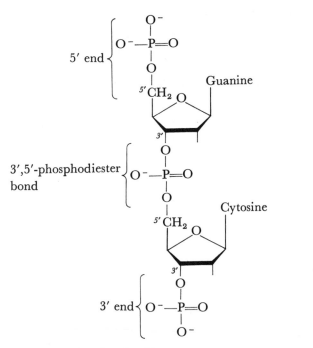

Structure of a dinucleotide

These equivalences suggested that the adenine of one DNA strand is always bound to thymine in the other strand and, similarly, that guanine is bound to cytosine. The proof of this base-pairing came when Watson and Crick proved by x-ray diffraction that the DNA structure was a double helix whose chains were complementary and antiparallel. By *complementary*, they meant that A binds to T and C to G between the chains. On the molecular level, this complementarity is accomplished by hydrogen bonding, as shown below. The chains are *antiparallel* because each end of the helix contains the 5′ end of one strand and the 3′ end of the other; hence, the chains travel in opposite directions.

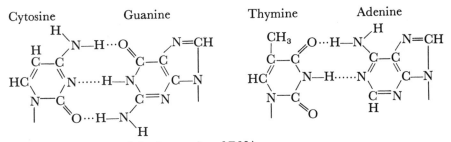

Hydrogen bonding of the base pairs of DNA

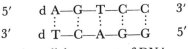

Antiparallel structure of DNA

Denaturation of DNA—i.e., the conversion of a double-stranded helix into single-stranded polynucleotides—occurs at high temperatures or the extremes of pH.

RNA and DNA are strongly acidic due to their many phosphoric acid residues. Basic proteins, such as histones, bind to DNA in the nucleus of the cell.

Viruses contain either RNA or DNA. Many of the DNA viruses have circular DNA.

Pancreatic ribonuclease (RNase) and deoxyribonuclease (DNase) hydrolyze phosphodiester bonds in dietary DNA and RNA to yield nucleotides and oligonucleotides (short-chain nucleotide polymers). Phosphodiesterase, found in various human tissues, also hydrolyzes phosphodiester bonds.

Problems 11–14

Match the bonds below to their descriptions in Problems 11–14:

A. Phosphodiester bond
B. Glycosidic bond

C. Phosphate ester
D. Hydrogen bonds

11. Links nucleoside to phosphoric acid.
12. Bridges the two DNA strands.
13. Joins the base to the pentose sugar.
14. Links nucleotides to one another in RNA.

Problem 15

Which deoxyoligonucleotide will base-pair in antiparallel sequence with dpACGGTACp? (Remember the convention that the 5′ end is written to the left of the 3′ end.)

A. dpACGGTACp
B. dpGTACCGTp
C. dpTGCCATGp

Problem 16

Which statement about the base content of DNA is *incorrect*?

A. A + T = G + C
B. A = T

C. G = C
D. A + G = C + T

Problem 17

Which coenzyme does *not* contain a nucleotide?

A. FAD
B. NAD+

C. CoA
D. CoQ

ANSWERS

1. This is a purine because it contains a pyrimidine ring fused to an imidazole ring. The pyrimidine ring has two hydroxyl groups and is in the enol or lactim tautomeric form. The imidazole group is also hydroxylated. This is the enol tautomer of uric acid.

2. This is a pyridine ring, not a pyrimidine ring. Pyridine has a single nitrogen atom, whereas pyrimidine has two. There is a carboxyl group attached to the pyridine ring. This is nicotinic acid, or niacin, a B vitamin.

3. This pyrimidine has two keto groups (the lactam form) plus a fluoride group. It is 5-fluorouracil, an analog of uracil. It interferes with rapid cellular proliferation and is used to treat certain cancers.

4. This is a pyridine ring rather than a pyrimidine. Attached to it are methyl, hydroxyl, aldehyde, and methanol groups. This is pyridoxal, a form of vitamin B_6.

5. This is the amino acid histidine, with its imidazole ring. It is not a purine, because it lacks a pyrimidine ring.

6. The left-hand ring is a pyrimidine ring (the enol or lactim tautomer) with an amino group. The right-hand ring is neither a pyrimidine nor an imidazole ring. Hence this compound is not a purine. The double rings constitute the pteridine nucleus of pteroylglutamic acid, or folic acid.

7. An aromatic ring is fused with a five-membered nitrogen-containing ring. This is an indole group, which identifies the compound as tryptophan.

8. C. (This is CDP.)

9. A.

10. B.

11. C.

12. D.

13. B.

14. A.

15. B.

16. A. The ratio of $(A + T)/(G + C)$ varies widely among different species but is rarely 1.0.

17. D.

REFERENCES

Barker, R. *Organic Chemistry of Biological Compounds*. Englewood Cliffs, N.J.: Prentice-Hall, 1971. Pp. 281–304.

Bhagavan, N. V. *Biochemistry—A Comprehensive Review*. Philadelphia: Lippincott, 1974. Pp. 297–308.

Lehninger, A. L. *Biochemistry: The Molecular Basis of Cell Structure and Function* (2nd ed.). New York: Worth, 1975. Pp. 309–324, 861–865.

Rich, A., and RajBhandary, U. Transfer RNA: molecular structure, sequence, and properties. *Annu. Rev. Biochem.* 45:805, 1976.

White, A., Handler, P., and Smith, E. L. *Principles of Biochemistry* (5th ed.). New York: McGraw-Hill, 1973. Pp. 180–202.

10 Carbohydrate Metabolism and Biosynthesis

You should plan to spend at least fifteen hours to learn this vitally important topic of carbohydrate metabolism and biosynthesis.

Dietary carbohydrate consists mainly of the polysaccharides, amylose and amylopectin, and the disaccharides, sucrose and lactose, with small amounts of free glucose and fructose. The principal monosaccharide derived from the intestinal hydrolysis of dietary carbohydrate is glucose. Fructose and galactose appear in lesser amounts, and they must be especially channeled into the mainstream of glucose breakdown, called *glycolysis* (lysis of glucose).

FRUCTOSE METABOLISM

Most human tissues cannot utilize fructose. The principal organs that metabolize fructose are the liver, kidneys, and intestine.

Theoretically, the simplest pathway would be to phosphorylate fructose to fructose-6-phosphate (fructose-6-P), an intermediate in glycolysis. Unfortunately, however, this reaction does not occur to any appreciable extent. Instead, fructokinase phosphorylates fructose to fructose-1-P. Fructose-1-P aldolase, in turn, splits this ketohexose-1-P into dihydroxyacetone phosphate (DHAP), a ketotriose-1-P, and glyceraldehyde, an aldotriose. DHAP and glyceraldehyde can then proceed through glycolysis to yield energy, or they can be converted to glucose, which can then be utilized by all the tissues of the body.

Hereditary fructokinase deficiency causes no symptoms, but affected individuals will spill fructose into their urine. This fructosuria might be misinterpreted as glucosuria, because both yield a positive reducing-sugar test.

Fructose-1-P aldolase deficiency, called *hereditary fructose intolerance*, causes fructose-1-P to accumulate inside liver cells after sucrose or fructose ingestion. This accumulation results in hypoglycemia (low serum glucose levels) and vomiting. The treatment of hereditary fructose intolerance is avoidance of dietary fructose and sucrose.

Some intravenous solutions contain fructose rather than glucose. Because fructose must be converted to glucose in the liver, kidneys, and intestine before it can be used by other organs, it is, in this respect, inferior to glucose. Fructose should generally not be used intravenously; however, it has been used successfully to speed up the breakdown of ethanol.

GALACTOSE METABOLISM

Like fructose, only a few organs can metabolize galactose; the most important of these are the liver and erythrocytes.

Galactose metabolism requires the formation of uridine diphosphate-glucose (UDP-glucose), which is created from glucose-1-P and UTP:

$$\text{Glucose-1-P} + \text{UTP} \rightleftharpoons \text{UDP-glucose} + \text{PP}_i$$

Galactokinase phosphorylates galactose to galactose-1-P. Next, a uridyl transferase removes glucose-1-P from UDP-glucose and creates UDP-galactose.

Humans have two uridyl transferases: galactose-1-P and hexose-1-P uridyl transferase. Finally, UDP-glucose epimerase changes the OH configuration at carbon 4' of UDP-galactose to yield UDP-glucose, an intermediate in glycogen synthesis:

$$\text{Galactose} + \text{ATP} \xrightleftharpoons{\text{galactokinase}} \text{galactose-1-P} + \text{ADP}$$

$$\text{Galactose-1-P} + \text{UDP-glucose} \xrightleftharpoons{\text{uridyl transferase}} \text{UDP-galactose} + \text{glucose-1-P}$$

$$\text{UDP-galactose} \xrightleftharpoons{\text{epimerase}} \text{UDP-glucose}$$

Galactosemia, an increased level of serum galactose, has either of two causes: hereditary galactokinase deficiency or hereditary galactose-1-P uridyl-transferase deficiency. In both disorders, galactose is reduced to its sugar alcohol, galactitol, which may be deposited inside the lens of the eye. Galactitol increases the osmotic pressure inside the lens and can cause cataracts. In only the uridyl-transferase deficiency is galactose-1-P trapped within the liver cells and erythrocytes, leading to hepatomegaly, impaired liver function, and mental retardation if the dietary galactose intake continues. (The uridyl-transferase deficiency resembles hereditary fructose intolerance. In both, a hexose-1-P cannot be metabolized nor can it leave the liver cells because of its charge; hence, it accumulates and causes hepatomegaly and liver dysfunction.) The treatment for both types of galactosemia is to eliminate milk products, the source of lactose, which contains galactose.

GLYCOLYSIS

Glycolysis is the main pathway for carbohydrate catabolism in virtually every human tissue (Fig. 10-1). (You should review glycolysis over and over again, recopying the reactions until you have memorized its entire sequence.) In studying glycolysis, particular focus should be placed on the two ATP-consuming steps, the two reactions that produce ATP, and the two oxidation-reduction reactions that involve NAD^+-NADH. The three irreversible steps—those catalyzed by hexokinase, phosphofructokinase, and pyruvate kinase—control the rate of glycolysis.

Every glycolytic intermediate between glucose and pyruvate contains phosphate. Because phosphate is highly ionized, these intermediates cannot leave the cell. Besides trapping these compounds, the phosphate groups are also used to phosphorylate ADP to ATP.

The enzymes of glycolysis are in the *cytoplasm* rather than in the mitochondria. Since glycolysis is essential, the systemic lack of one of the eleven glycolytic enzymes is incompatable with life. The physician may, however, encounter patients whose erythrocytes and sometimes leukocytes are deficient in one or another glycolytic enzyme; they develop hemolytic anemia.

To initiate glycolysis, *hexokinase irreversibly phosphorylates glucose to glucose-6-P,* consuming ATP. As in all phosphorylation reactions, Mg^{+2} is an essential cofactor. The hexokinase reaction is one of the three major rate-controlling steps of glycolysis. The liver possesses an additional kinase, glucokinase, with a

Figure 10-1

Glycolysis. Single arrows denote reactions that are essentially irreversible, while double arrows represent reversible reactions.

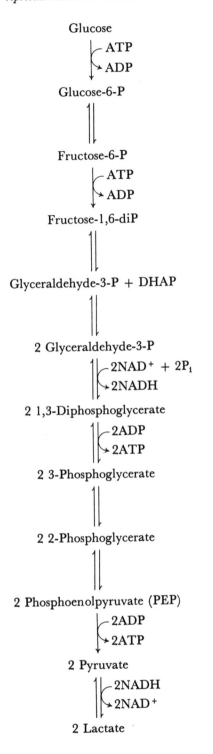

higher Michaelis constant (K_m) than hexokinase to handle large surges of glucose after meals.

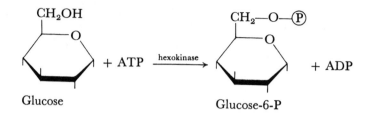

Glucose + ATP → (hexokinase) Glucose-6-P + ADP

The second step of glycolysis is *the reversible isomerization of glucopyranose-6-P to fructofuranose-6-P by glucose-phosphate isomerase*:

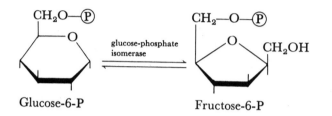

Glucose-6-P ⇌ (glucose-phosphate isomerase) Fructose-6-P

Phosphofructokinase (PFK) then irreversibly phosphorylates fructose-6-P to fructose-1,6-diP. This rate-limiting step in glycolysis is under allosteric control (which is explained in the subsequent section on gluconeogenesis). Up to this point, for every mole of fructose-1,6-diP produced from glucose, two moles of ATP are consumed.

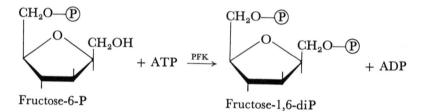

Fructose-6-P + ATP → (PFK) Fructose-1,6-diP + ADP

The next reaction marks the transition between the hexose and triose stages of glycolysis. *Fructose-diphosphate aldolase reversibly cleaves the bond between carbons 3 and 4 of fructose-1,6-diP to create two triose phosphates: dihydroxyacetone phosphate (DHAP) and glyceraldehyde-3-P*. The carbon atoms of fructose are numbered to show the fate of radioactive labels observed through this step:

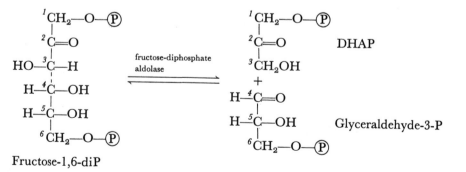

Fructose-1,6-diP ⇌ (fructose-diphosphate aldolase) DHAP + Glyceraldehyde-3-P

DHAP can proceed no farther in glycolysis until its keto group is reversibly transformed into an aldehyde group by triose-phosphate isomerase to create another molecule of glyceraldehyde-3-P. Note that as a result of this reaction, a radioactive label on either carbon 1 or carbon 6 of glucose will appear at the phosphorylated carbon of glyceraldehyde-3-P. Similarly, a label at carbon 2 or 5 of glucose appears on the middle carbon of glyceraldehyde-3-P, whereas a label at carbon 3 or 4 appears on its aldehyde group.

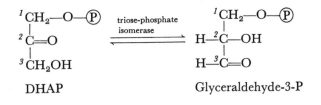

DHAP   Glyceraldehyde-3-P

Finally, we arrive at the first energy-producing reaction of glycolysis, *the reversible glyceraldehyde-phosphate dehydrogenase reaction.* Two processes occur in this reaction:

1. The aldehyde group of glyceraldehyde-3-P is oxidized to a carboxylic acid group, thereby reducing NAD^+ to NADH.
2. Inorganic phosphate, P_i, joins to the OH group of this carboxyl group, creating 1,3-diphosphoglycerate. Thus, ATP is not consumed in this phosphorylation.

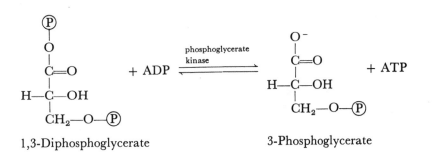

Glyceraldehyde-3-P   1,3-Diphosphoglycerate

In erythrocytes, 1,3-diphosphoglycerate is converted to 2,3-diphosphoglycerate (2,3-DPG); the latter shifts the O_2 dissociation curve of hemoglobin to the right (p. 15).

The next step in glycolysis is *the removal of a phosphate from 1,3-diphosphoglycerate by phosphoglycerate kinase to yield 3-phosphoglycerate plus ATP.* This enzyme is named for the reverse reaction.

1,3-Diphosphoglycerate   3-Phosphoglycerate

Next, phosphoglyceromutase shifts the phosphate from carbon 3 to carbon 2, yielding 2-phosphoglycerate:

3-Phosphoglycerate   2-Phosphoglycerate

Enolase then dehydrates 2-phosphoglycerate to create phosphoenolpyruvate (PEP), an enol, i.e., a compound with the $-C=C(OH)-$ structure:

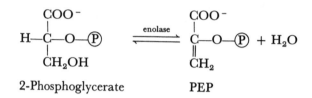

2-Phosphoglycerate   PEP

Because PEP has a very high energy phosphate group (-14.8 kcal/mole), its hydrolysis easily drives ATP production, which requires 7.3 kcal/mole. *Pyruvate kinase catalyzes this irreversible phosphate transfer from PEP to ADP to yield pyruvate:*

PEP   Pyruvate

Pyruvate kinase is an allosteric enzyme inhibited by ATP, and it, too, is named for the reverse reaction.

Pyruvate is a key link in metabolism that ties together glycolysis, the tricarboxylic acid (TCA) cycle, amino acid metabolism, and fatty acid oxidation.

In tissues with adequate oxygen supply, pyruvate formation is the last step of glycolysis. Most of this pyruvate is then oxidized and decarboxylated to form an acetyl group, which, when combined with coenzyme A (acetyl-CoA), enters the tricarboxylic-acid cycle.

Anaerobic tissues, such as exercising muscle or, in cases of coronary-artery disease, an inadequately perfused myocardium, cannot utilize the TCA cycle or oxidative phosphorylation. Hence, NADH produced by the glyceraldehyde-phosphate dehydrogenase reaction accumulates, while NAD^+ becomes scarce. This scarcity of NAD^+ would stop glycolysis were it not for the lactate dehydrogenase (LDH) reaction, which uses NADH to reduce pyruvate to lactate plus NAD^+. *Anaerobic glycolysis,* therefore, neither produces nor consumes NADH.

Phenformin, an oral hypoglycemic drug (i.e., one that lowers blood glucose levels) simulates anaerobic glycolysis and increases lactic-acid production.

$$\underset{\text{Pyruvate}}{\begin{matrix} \text{COO}^- \\ | \\ \text{C}{=}\text{O} \\ | \\ \text{CH}_3 \end{matrix}} + \text{NADH} + \text{H}^+ \underset{\text{LDH}}{\rightleftharpoons} \underset{\text{Lactate}}{\begin{matrix} \text{COO}^- \\ | \\ \text{H}{-}\text{C}{-}\text{OH} \\ | \\ \text{CH}_3 \end{matrix}} + \text{NAD}^+$$

Lactic acid produced in anaerobic tissues diffuses into the bloodstream and reaches the liver, where it is oxidized back to pyruvate and then metabolized aerobically.

In lactic acidosis, a common metabolic disorder, lactate is produced so rapidly that the liver cannot remove it fast enough. The blood lactic-acid level rises, and metabolic acidosis ensues.

ENERGETICS OF GLYCOLYSIS

The overall reaction of glycolysis alone, using either free glucose, fructose, or galactose and yielding pyruvate, creates two moles of NADH and four moles of ATP but consumes two moles of ATP per mole of hexose; the net gain is two moles each of ATP and NADH.

Since glycolysis occurs only in the cytoplasm, the NADH produced must be transported into the mitochondria so that it can undergo oxidative phosphorylation. Since neither NADH nor NAD^+ can penetrate the mitochondrial membranes, shunts must transport this reducing power into the mitochondria. Two such shunts, discussed in Chapter 11, are the glycerol-phosphate and the malate shuttle, which yield three and two moles ATP, respectively, per mole of NADH transferred. Since there is still controversy about which shuttle dominates, the total net ATP gain in aerobic glycolysis is either six or eight moles ATP per mole of hexose oxidized to pyruvate. The further oxidation of pyruvate to CO_2 and H_2O in the TCA cycle yields either 36 or 38 moles ATP per mole of hexose oxidized to CO_2 and H_2O.

In anaerobic glycolysis, on the other hand, only two moles of ATP and no NADH are produced per mole of hexose. Anaerobic tissues metabolize glucose much more rapidly than aerobic tissues do to compensate for this meager energy gain. The overall reaction of anaerobic glycolysis may thus be written:

$$\text{Glucose} + 2\text{ADP} + 2\text{P}_i \longrightarrow 2\text{ lactate} + 2\text{ATP} + 2\text{H}_2\text{O}$$

Problems 1–4

Name the structures and enzymes in the reactions below. Which reactions are irreversible?

1.
$$\underset{\text{CH}_2}{\overset{\text{COO}^-}{\underset{|}{\overset{|}{\text{C}{-}\text{O}{-}\textcircled{P}}}}} + \text{ADP} \longrightarrow \underset{\text{CH}_3}{\overset{\text{COO}^-}{\underset{|}{\overset{|}{\text{C}{=}\text{O}}}}} + \text{ATP}$$

2. UDP-galactose $\longrightarrow$ UDP-glucose

3.

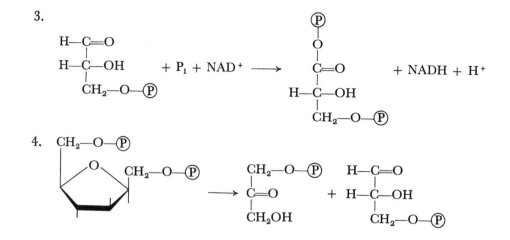

4.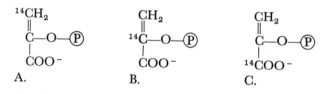

Problem 5

If glucose were labeled with ^{14}C at carbon 3, where would the label appear in PEP?

$$^{14}CH_2$$
$$\| $$
$$C-O-\textcircled{P}$$
$$|$$
$$COO^-$$
A.

$$CH_2$$
$$\|$$
$$^{14}C-O-\textcircled{P}$$
$$|$$
$$COO^-$$
B.

$$CH_2$$
$$\|$$
$$C-O-\textcircled{P}$$
$$|$$
$$^{14}COO^-$$
C.

The diagram below will refresh your memory of the fate of each carbon atom in glycolysis:

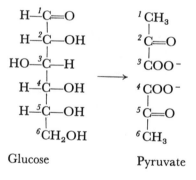

Glucose   Pyruvate

Problem 6

If glucose were labeled with ^{14}C at carbon 5, where would the label appear in 3-phosphoglycerate?

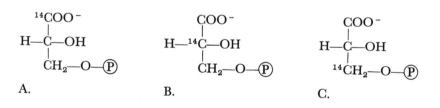

$$^{14}COO^-$$
$$|$$
$$H-C-OH$$
$$|$$
$$CH_2-O-\textcircled{P}$$
A.

$$COO^-$$
$$|$$
$$H-^{14}C-OH$$
$$|$$
$$CH_2-O-\textcircled{P}$$
B.

$$COO^-$$
$$|$$
$$H-C-OH$$
$$|$$
$$^{14}CH_2-O-\textcircled{P}$$
C.

Problem 7

Infants with hereditary galactokinase deficiency will develop cataracts if fed lactose, because:

A. galactose-1-P accumulates.
B. galactose is reduced to galactitol, which causes water retention inside the lens.
C. galactose itself causes the lens to retain water.

Problem 8

The net reaction of aerobic glycolysis of one mole of glucose to two moles pyruvate yields:

A. $4ATP + 2NADH + 2H^+$
B. $2ATP + NADH + H^+$
C. $2ATP + 2NADH + 2H^+$
D. $2ATP$

Problem 9

Which reaction catalyzed by the enzymes below is *not* rate-controlling in glycolysis?

A. Pyruvate kinase
B. PFK
C. Hexokinase
D. Triose-phosphate isomerase

GLUCONEOGENESIS

Gluconeogenesis literally means the new formation of glucose; it involves the conversion of three- and four-carbon compounds to the six-carbon compound glucose. The fuels for gluconeogenesis are the sugar alcohol, glycerol, and noncarbohydrates, such as lactate and the α-keto acids, pyruvate and oxaloacetate.

The brain and exercising skeletal muscle require glucose as their principal fuel. During fasting, the liver stores only enough glycogen to supply the body with glucose for 12 to 24 hours. The sole source of glucose in prolonged fasting is that supplied by gluconeogenesis from glycerol (derived from triglyceride hydrolysis) and from α-keto acids (derived from amino-acid catabolism); i.e., the glycerol backbone of triglycerides and the protein of skeletal muscle are consumed to provide glucose for the brain and exercising skeletal muscle.

Gluconeogenesis is confined almost entirely to the liver, kidneys, and intestinal epithelium; certain enzymes required in gluconeogenesis are present only in these three organs.

The central pathway of gluconeogenesis from the α-keto acids is the conversion of pyruvate to glucose, a net reversal of glycolysis. Of the 11 reactions of glycolysis, eight are reversible and can, therefore, be used in gluconeogenesis. There are three irreversible reactions of glycolysis, however, that must be bypassed in gluconeogenesis: the pyruvate-kinase, PFK, and hexokinase reactions.

Two enzymes are required to bypass the pyruvate-kinase reaction.

Pyruvate carboxylase, using biotin as a coenzyme, carboxylates pyruvate to produce oxaloacetate (OAA or $^-OOCC(=O)CH_2COO^-$). Almost all carboxylation of α-keto acids consume ATP and require biotin. Pyruvate carboxylase is an allosteric enzyme that requires the presence of acetyl-CoA. By supplying OAA, this pyruvate-carboxylase reaction is the principal means used to replenish the TCA cycle.

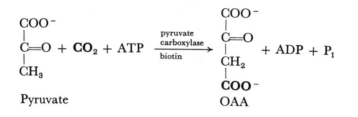

The second enzyme in this bypass is *PEP carboxykinase*, which phosphorylates and decarboxylates OAA to create PEP. This reaction requires GTP, which is obtained by phosphorylating GDP with ATP. As shown below, one mole of GTP has the same energy value as one mole of ATP:

$$ATP + GDP \rightleftharpoons ADP + GTP$$

Thus, two moles of ATP are consumed in bypassing the pyruvate-kinase reaction.

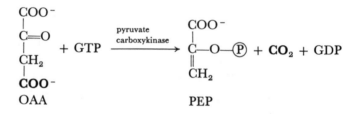

Once formed, PEP proceeds in the reverse direction through the steps of glycolysis until it reaches the next irreversible step: the PFK reaction. Gluconeogenesis uses *hexosediphosphatase* to remove the phosphate group from carbon 1 of fructose. This reaction does not regenerate the ATP that would be consumed in the PFK reaction:

$$\text{Fructose-1,6-diP} + H_2O \xrightarrow{\text{hexosediphosphatase}} \text{fructose-6-P} + P_i$$

If both the PFK and hexosediphosphatase reactions occurred simultaneously, this would result in a futile cycle that consumed ATP. Fortunately, both enzymes are allosterically controlled to prevent a futile cycle. If a cell abounds in "energy," then ATP, NADH, and citrate will be plentiful, while ADP and AMP will be scarce. Citrate stimulates hexosediphosphatase, thereby augmenting gluconeogenesis; ATP, NADH, and citrate inhibit PFK, thereby slowing glycolysis. On the other hand, if a cell is low in "energy," ADP and AMP will abound, while ATP, NADH, and citrate will be scarce. ADP and AMP stimulate PFK, thereby augmenting glycolysis. Meanwhile, AMP inhibits hexosediphosphatase to slow gluconeogenesis.

The next step of gluconeogenesis is the reversal of a glycolysis reaction: glucose-phosphate isomerase converts fructose-6-P to glucose-6-P.

Gluconeogenesis must then bypass the third irreversible step of glycolysis: the hexokinase reaction. *Glucose-6-phosphatase* removes phosphate from glucose-6-P to yield the end product, glucose. This reaction, which also yields inorganic phosphate, again fails to recover the ATP that would be consumed in the hexokinase reaction:

$$\text{Glucose-6-P} + \text{H}_2\text{O} \xrightarrow{\text{glucose-6-phosphatase}} \text{glucose} + \text{P}_i$$

Type I glycogen storage disease results from the hereditary absence of glucose-6-phosphatase. Since this prevents the final step of gluconeogenesis, the liver and kidneys are unable to release glucose into the blood. Such individuals must eat every 3 to 5 hours to prevent fasting hypoglycemia.

The overall reaction for converting pyruvate to glucose can be summarized as follows:

$$2 \text{ Pyruvate} + 2\text{ATP} + 2\text{GTP} \rightarrow \rightarrow 2\text{PEP} + 2\text{ADP} + 2\text{GDP} + 2\text{P}_i$$

$$2\text{ATP} + 2\text{GDP} \longrightarrow 2\text{ADP} + 2\text{GTP}$$

$$2\text{PEP} + 2\text{ATP} \rightarrow \rightarrow \rightarrow 2 \text{ 1,3-diphosphoglycerate} + 2\text{ADP}$$

$$2 \text{ 1,3-diphosphoglycerate} + 2\text{NADH} + 2\text{H}^+ \longrightarrow$$
$$2 \text{ glyceraldehyde-3-P} + 2\text{NAD}^+ + 2\text{P}_i$$

$$2 \text{ Glyceraldehyde-3-P} + 2\text{H}_2\text{O} \rightarrow \rightarrow \rightarrow \rightarrow \text{glucose} + 2\text{P}_i$$

Net: $2 \text{ Pyruvate} + 6\text{ATP} + 2\text{NADH} + 2\text{H}^+ + 2\text{H}_2\text{O} \longrightarrow$
$$\text{glucose} + 6\text{ADP} + 6\text{P}_i + 2\text{NAD}^+$$

Thus, gluconeogenesis from pyruvate consumes six moles ATP and two moles NADH. The reverse process—i.e., glycolysis to pyruvate—produces two moles each of ATP and NADH.

Gluconeogenesis from the glycerol (derived from triglyceride hydrolysis) begins by the phosphorylation of glycerol to glycerol-P and then the oxidation of this to DHAP. DHAP then proceeds in a reverse direction through the steps of glycolysis; hexosediphosphatase and glucose-6-phosphatase are again employed to bypass the two irreversible reactions of glycolysis.

An amino acid is termed *glucogenic* or *glycogenic* if it can be converted to glucose via gluconeogenesis. To enter gluconeogenesis, an amino acid must be broken down to pyruvate, 3-phosphoglycerate, or an intermediate in the tricarboxylic-acid cycle. Sufficient dietary carbohydrate results in protein sparing, because amino-acid breakdown for gluconeogenesis is reduced.

HEXOSE-MONOPHOSPHATE SHUNT

The *hexose-monophosphate (HMP) shunt* derives its name from the initial reactant of this pathway: glucose-6-P. It is also termed the phosphogluconate pathway, because 6-phosphogluconate is one of its intermediates. A third name is the pentose-phosphate pathway, so called because ribose-5-P is one of its products. The main purposes of the hexose-monophosphate shunt are:

1. to produce ribose-5-P for nucleotide synthesis,
2. to produce NADPH from $NADP^+$ for fatty-acid and steroid biosynthesis and for maintaining reduced glutathione inside erythrocytes, and
3. to interconvert pentoses and hexoses.

Organs that actively synthesize fatty acids and steroids—such as the lactating mammary gland, the liver, the adrenal cortex, and adipose tissue—channel a significant proportion of their glucose into the HMP shunt. At least 30% of hepatic glucose enters this pentose-phosphate pathway.

In the initial reaction of the HMP pathway, glucose-6-P dehydrogenase (G-6-PDH) oxidizes glucose-6-P to 6-phosphogluconolactone, producing NADPH:

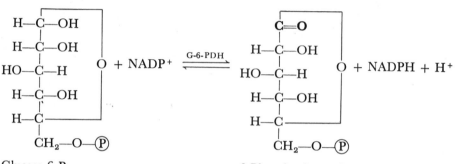

Glucose-6-P   6-Phosphogluconolactone

The inherited deficiency of glucose-6-P dehydrogenase in erythrocytes is one of the world's most common enzyme deficiency diseases, particularly among Mediterranean peoples and blacks. In the United States, 10% of black males are homozygous for this deficiency disease in their erythrocytes. G-6-PDH deficiency stops the HMP shunt, thereby preventing NADPH production. Erythrocytes need NADPH to maintain glutathione, a tripeptide, in its reduced form to protect them from oxidizing agents. Individuals with this deficiency develop hemolytic anemia after receiving oxidizing drugs or eating fava beans (hence the name "favism" for this disease).

Subsequent reactions of the HMP pathway decarboxylate the hexose sugars to form pentoses, such as ribose-5-P, an essential component of nucleic acids. Other reactions rearrange triose, tetrose, pentose, hexose, and heptulose sugars. The net overall reaction is:

$$6 \text{ Glucose-6-P} + 12NADP^+ + 7H_2O \longrightarrow$$
$$5 \text{ glucose-6-P} + 6CO_2 + 12NADPH + P_i + 12H^+$$

Unlike glycolysis, this shunt can oxidize glucose-6-P completely to CO_2 and H_2O.

In this pathway, there is a transketolase reaction that requires TPP as its coenzyme. In thiamine deficiency, the transketolase activity in erythrocytes is reduced.

Problems 10–13

Match the allosteric modulators below to their effect on PFK and hexosediphosphatase:

A. AMP
B. 3′,5′-cyclic AMP
C. ATP
D. ADP
E. citrate

10. Positive effector(s) for PFK
11. Allosterically inhibits PFK
12. Positive modulator(s) for hexosediphosphatase
13. Allosterically inhibits hexosediphosphatase

Problem 14

During cardiac catheterization, the lactate level in the coronary sinus can be measured to help diagnose myocardial ischemia (the result of inadequate O_2 delivery) in coronary-artery disease. The myocardium normally consumes lactic acid, so that the coronary-sinus lactate level is lower than the peripheral venous lactate level. Ischemic myocardial tissue, however, produces lactate. Choose the *false* statement:

A. Normal myocardium oxidizes lactate to pyruvate and feeds this pyruvate into the TCA cycle.
B. Ischemic myocardium cannot fully utilize the TCA cycle.
C. Normal myocardium oxidizes lactate to pyruvate and then converts most of this pyruvate to glucose.
D. Ischemic myocardium oxidizes glucose more rapidly than normal myocardium and overproduces lactate.

Problems 15–21

Match the enzymes below to the reactions given in Problems 15–21:

A. Pyruvate carboxylase
B. PEP carboxykinase
C. Pyruvate kinase
D. PFK
E. Hexosediphosphatase
F. Hexokinase
G. Glucose-6-phosphatase

15. Fructose-6-P + ATP $\longrightarrow$ fructose-1,6-diP + ADP

16. Pyruvate + CO_2 + ATP $\xrightarrow{\text{biotin}}$ OAA + ADP + P_i

17. Glucose-6-P + H_2O $\longrightarrow$ glucose + P_i

18. PEP + ADP $\longrightarrow$ pyruvate + ATP

19. Fructose-1,6-diP + H_2O $\longrightarrow$ fructose-6-P + P_i

20. Glucose + ATP $\longrightarrow$ glucose-6-P + ADP

21. OAA + GTP $\longrightarrow$ PEP + CO_2 + GDP

Match the enzymes below with the features of their deficiency diseases given in Problems 22–25:

A. Glucose-6-P dehydrogenase
B. Galactose-1-P uridyl transferase
C. Glucose-6-phosphatase
D. Galactokinase
E. Fructokinase
F. Fructose-1-P aldolase

22. Hypoglycemia and vomiting after sucrose ingestion.
23. Hepatomegaly, liver dysfunction, cataracts, and mental retardation following prolonged lactose ingestion.
24. Fasting hypoglycemia; glycogen accumulates excessively in the liver.
25. Insufficient reduced glutathione inside erythrocytes predisposes to hemolytic anemia.

GLYCOGENOLYSIS

Glycogenolysis means the lysis or breakdown of glycogen. The first step of glycogen breakdown is the phosphorylation of its α-1,4 glycosidic bonds by glycogen phosphorylase. *Glycogen phosphorylase* exists in two forms: phosphorylase b, an inactive dimer, and phosphorylase a, an active tetramer with an additional phosphate group. As shown in Figure 10-2, epinephrine or

Figure 10-2

Cyclic AMP-mediated activation of glycogen phosphorylase and inhibition of glycogen synthase.

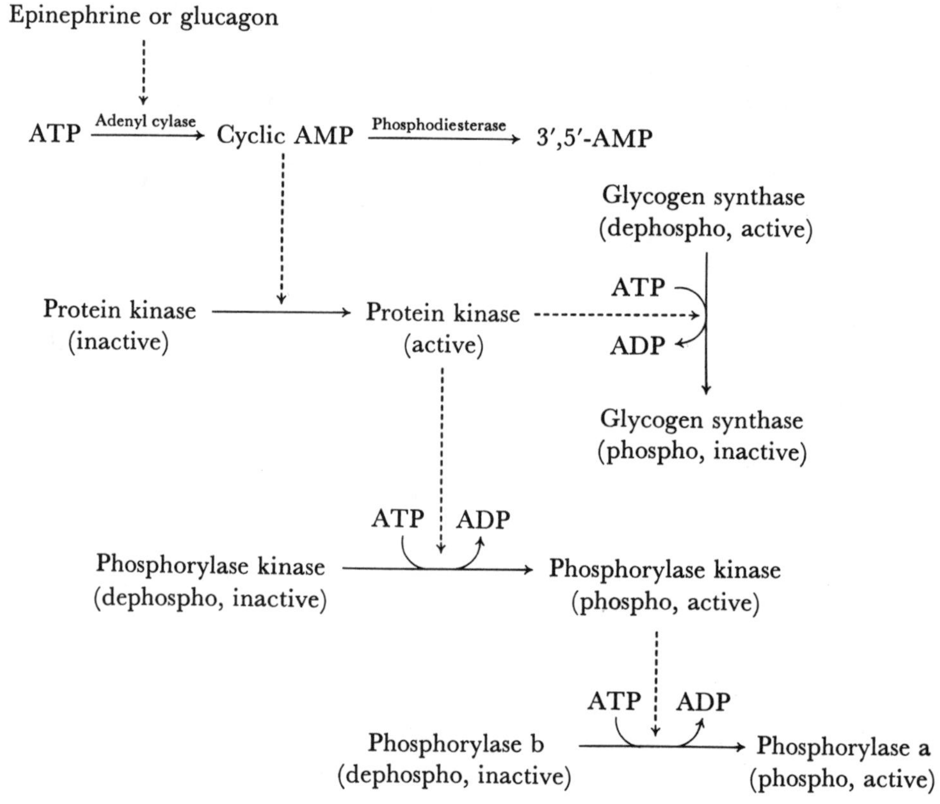

glucagon stimulates adenyl cyclase in the membranes of liver cells to convert ATP to cyclic AMP. Cyclic AMP then converts inactive protein kinase to active protein kinase, which, in turn, adds phosphate to phosphorylase kinase to yield active (phospho) phosphorylase kinase. This enzyme then adds phosphate to glycogen phosphorylase b to yield active (phospho) glycogen phosphorylase a.

In conjunction with activating glycogenolysis, epinephrine and glucagon inhibit glycogen synthesis. The active protein kinase that is formed as a result of stimulation by epinephrine or glucagon adds phosphate to the active (dephospho) form of *glycogen synthase* to yield inactive (phospho) glycogen synthase. Thus, glycogenolysis is usually prevented from operating simultaneously with glycogen synthesis, thereby avoiding a futile cycle.

In addition to activation by covalent phosphate binding, both glycogen phosphorylase and glycogen synthase are also allosterically regulated. AMP allosterically stimulates glycogen phosphorylase b, and glucose-6-P stimulates glycogen synthase.

Cyclic-AMP phosphodiesterase inactivates cyclic AMP. The primary agents in the treatment of bronchial asthma, the theophylline drugs, act to dilate bronchi by inhibiting cyclic-AMP phosphodiesterase.

In the process of glycogenolysis, glycogen phosphorylase a yields a limit dextrin after it has released successive glucose-1-P residues from the branch chains of the glycogen. Other enzymes, such as the debranching enzyme amylo-1,6-glucosidase, must then cleave the α-1,6 linkages.

The hereditary absence of muscle glycogen phosphorylase, termed type V glycogen storage disease, causes normal glycogen to accumulate in skeletal muscle, which produces muscle weakness.

Type III glycogen storage disease, the inherited deficiency of amylo-1,6-glucosidase in the liver and in heart and skeletal muscle, causes limit dextrins to accumulate in these tissues.

Phosphoglucomutase converts the glucose-1-P liberated in glycogenolysis to glucose-6-P, which may have one of three fates: it can proceed through glycolysis, it can enter the HMP shunt, or, in the liver, kidneys, and intestinal epithelium, it can be cleaved by glucose-6-phosphatase to form free glucose and inorganic phosphate.

GLYCOGENESIS

Glycogenesis means glycogen synthesis. The initial step of glycogenesis is the conversion of glucose-6-P to glucose-1-P by phosphoglucomutase. Next, glucose-1-P binds to UTP to create UDP-glucose. This reaction resembles the binding of galactose to UTP to form UDP-galactose.

$$\text{Glucose-1-P} + \text{UTP} \longrightarrow \text{UDP-glucose} + \text{PP}_i$$

Glycogen synthase then adds this bound glucose in α-1,4 linkage to the glycogen polymer, liberating UDP:

$$\text{UDP-glucose} + (\text{glucose})_n \longrightarrow (\text{glucose})_{n+1} + \text{UDP}$$

For every glucose molecule incorporated into glycogen in glycogenesis, one mole of ATP is expended to produce glucose-6-P from glucose and one mole of UTP is spent to create UDP-glucose. Following glycogenolysis, the one mole of ATP spent in glycogen synthesis is recovered when the glucose-1-P produced undergoes glycolysis; in fact, three moles of ATP are generated by glycolysis per mole of glucose-1-P removed from glycogen.

In glycogenesis, an α-1,4 glucan branching enzyme removes α-1,4-linked glucose oligosaccharides and reattaches them by α-1,6 bonds to create the proper branching. In type IV glycogen storage disease, the hereditary absence of this branching enzyme leads to the accumulation of long glucose polymers with few branches.

The process of glycogenesis is coupled somehow to the influx of K^+ into the cells. Hyperkalemia (a high serum K^+ level) is usually treated initially by giving glucose and insulin to induce glycogenesis and the concomitant removal of K^+ from the serum.

HORMONAL CONTROL OF CARBOHYDRATE METABOLISM

Glucose does not readily penetrate through the cell membranes of most human tissues. To facilitate glucose entry into cells, insulin must be present. The brain, liver, kidneys, and blood cells, however, do not need insulin for glucose transport. The rapid intravenous injection of 50 units of regular insulin, for instance, will lower the serum glucose level rapidly and cause insulin shock, which is characterized by impaired consciousness or coma, sweating, anxiety, and various neurologic abnormalities that result from an inadequate glucose supply to the brain. Hypoglycemic shock is treated by administering intravenous glucose.

Insulin also promotes amino-acid uptake by cells and stimulates protein synthesis, thereby reducing the amino-acid supply available for gluconeogenesis. By slowing gluconeogenesis, insulin again helps to lower the serum glucose level.

In addition, insulin promotes glycogenesis by stimulating glycogen synthase. Insulin does not, however, act via the second messenger, cyclic AMP. It probably makes protein kinase less sensitive to cyclic AMP, thereby blunting the glycogenolytic response to epinephrine or glucagon. Insulin is the only hormone that acts to lower serum glucose levels as well as to promote glucose storage.

Somatotrophic hormone (STH), also termed growth hormone (GH), antagonizes insulin activity by an unknown mechanism. Thus, STH is a glucose-mobilizing hormone; it raises the serum glucose level. Hyperglycemia is a feature of acromegaly, which is caused by a STH-producing pituitary tumor.

Thyroid hormones also have some effect in raising serum glucose levels. The mechanism for this is uncertain.

Glucocorticoids (the 11-hydroxy, C21 adrenocortical steroids) promote hyperglycemia by stimulating gluconeogenesis in two ways. First they promote protein and amino-acid breakdown, thereby providing more pyruvate and OAA as fuel for gluconeogenesis. In addition, they stimulate the liver to produce more gluconeogenetic enzymes, such as glucose-6-phosphatase.

Epinephrine and glucagon are also glucose-mobilizing hormones. They raise the serum glucose level mainly by stimulating glycogenolysis via cyclic

AMP in the liver and kidneys, as well as by inhibiting glycogenesis (see Fig. 10-2).

Diabetes mellitus has traditionally been considered as a deficiency of effective insulin action. More recently, however, it has been suggested that the inappropriately high serum glucagon levels in diabetes mellitus contribute to the hyperglycemia of diabetes mellitus.

The increase in serum glucose levels after a carbohydrate meal triggers insulin release while inhibiting the glucose-mobilizing hormones such as epinephrine, glucagon, and the glucocorticoids. Several hours after a meal, insulin secretion diminishes, and the levels of the glucose-mobilizing hormones increase. During fasting, virtually no insulin is detectable in the serum. Glucagon and epinephrine can maintain the serum glucose concentration for only 6 to 24 hours. After the hepatic glycogen reserves are depleted, glucocorticoids assume the dominant role of providing glucose via gluconeogenesis.

Problems 26–28

Match the features of glycogen storage diseases below to their consequences in Problems 26–28:

    A. Absent glycogen phosphorylase in muscles
    B. Amylo-1,6-glucosidase deficiency
    C. Absent glycogen synthase in liver
    D. Glucose-6-phosphatase deficiency in liver, kidneys, and intestine

26. Reduced amount of liver glycogen.
27. Limit dextrins accumulate.
28. Glycogen with normal structure accumulates in target organs.

Problem 29

Which statement about the control of glycogenolysis and glycogenesis is *incorrect*?

    A. Phosphorylase kinase receives a phosphate from ATP to become activated.
    B. ATP adds phosphate to glycogen phosphorylase b to activate it.
    C. Active protein kinase stimulates glycogen synthase.
    D. Cyclic AMP acts as a second messenger for epinephrine and glucagon.

Problems 30–35

Match the hormones below to their actions on carbohydrate metabolism given in Problems 30–35. You may choose one or more answers for each question.

    A. Insulin
    B. Glucocorticoids
    C. Somatotrophic hormone (STH)
    D. Glucagon
    E. Epinephrine
    F. Thyroid hormones

30. Hyperglycemic effect
31. Hypoglycemic action

32. Powerful stimulant of gluconeogenesis
33. Inhibits gluconeogenesis
34. Cause instantaneous glycogenolysis to counteract hypoglycemia
35. Stimulates glycogen synthase

Problem 36

All but one of the disorders below causes hypoglycemia. Which one produces hyperglycemia?

A. Insulin-producing pancreatic tumor
B. Glucose-6-phosphatase deficiency
C. Hypopituitarism (leads to reduced STH, thyroid hormones, and glucocorticoids)
D. Pheochromocytoma (adrenal tumor that causes overproduction of epinephrine)

Problem 37

How many moles of NTP (nucleoside triphosphate) are consumed when one glucose molecule is incorporated into glycogen and then removed and reconverted to glucose in the liver?

A. 0   C. 2
B. 1   D. 3

Problem 38

Account for each of the six moles of ATP and two moles of NADH required in the synthesis of one glucose molecule from two moles pyruvate.

ANSWERS

1. PEP + ADP $\longrightarrow$ pyruvate + ATP

   Irreversible (a rate-controlling step of glycolysis). Catalyzed by pyruvate kinase.

2. The configuration of the OH group at carbon 4 of galactose is reversed to produce glucose, an epimer of galactose. Catalyzed by an epimerase (UDP-glucose epimerase). Reversible.

3. Glyceraldehyde-3-P + P_i + NAD^+ $\rightleftharpoons$ 1,3-diphosphoglycerate + NADH

   Reversibly catalyzed by glyceraldehyde-phosphate dehydrogenase.

4. Fructose-1,6-diP $\rightleftharpoons$ DHAP + glyceraldehyde-3-P

   Reversibly catalyzed by fructose-diphosphate aldolase.

5. C.
6. B.
7. B. (In galactokinase deficiency, no galactose-1-P can be formed.)
8. C.
9. D.
10. A, D.
11. C, E.
12. E. (ATP was formerly considered a positive modulator for this enzyme, but no longer.)
13. A.

14. C.
15. D.
16. A.
17. G.
18. C.
19. E.
20. F.
21. B.
22. F.
23. B.
24. C.
25. A.
26. C. (This glycogen storage disease differs from most other types in that it results in less tissue glycogen, rather than an accumulation of tissue glycogen.)
27. B. (Type III glycogen storage disease.)
28. A, D. (Types V and I, respectively, of glycogen storage disease.)
29. C. (Active protein kinase adds phosphate to the active glycogen synthase, thereby inactivating it. This is opposite to the effect of phosphate binding on the activation of glycogen phosphorylase.)
30. B, C, D, E, F.
31. A.
32. B.
33. A.
34. D, E.
35. A.
36. D.
37. C. (One ATP plus one UTP.)
38. Refer to the section on gluconeogenesis.

REFERENCES

Goodman, L. S., and Gilman, A. *The Pharmacological Basis of Therapeutics* (5th ed.). New York: MacMillan, 1975. Pp. 1378 (STH), 1481 (glucocorticoids), 1512–1516 (insulin), 1528 (glucagon).

Howell, R. The Glycogen Storage Diseases. In Stanbury, J., Wyngaarden, J., and Fredrickson, D. (Eds.), *The Metabolic Basis of Inherited Disease* (3rd ed.). New York: McGraw-Hill, 1972.

Lehninger, A. L. *Biochemistry: The Molecular Basis of Cell Structure and Function* (2nd ed.). New York: Worth, 1975. Pp. 417–439, 467–472, 623–632, 642–648.

Neufeld, E., Lim, T., and Shapiro, L. Inherited disorders of lysosomal metabolism. *Annu. Rev. Biochem.* 44:357, 1975.

Segal, S. Disorders of Galactose Metabolism. In Stanbury, J., Wyngaarden, J., and Fredrickson, D. (Eds.), *The Metabolic Basis of Inherited Disease* (3rd ed.). New York: McGraw-Hill, 1972.

Sherwin, R., Fisher, M., Hendler, R., and Felig, P. Hyperglucagonemia and blood glucose regulation in normal, obese, and diabetic subjects. *N. Engl. J. Med.* 294:455, 1976.

White, A., Handler, P., and Smith, E. L. *Principles of Biochemistry* (5th ed.). New York: McGraw-Hill, 1973. Pp. 418–460, 466–484, 488–490.

11 Tricarboxylic Acid Cycle and Oxidative Phosphorylation

To learn these crucial areas of biochemistry, you should spend at least fifteen hours on this unit.

Unlike carbohydrate metabolism, which takes place in the cytoplasm, the tricarboxylic acid (TCA) cycle and oxidative phosphorylation occur within the mitochondria.

THE TRICARBOXYLIC ACID CYCLE

The *tricarboxylic acid cycle*, also called the citric acid or Krebs cycle, is so named because several of its intermediates have three carboxyl groups: citrate, *cis*-aconitate, and isocitrate. The remaining six intermediates are dicarboxylic acids.

The TCA cycle is the central hub in the metabolism of carbohydrates, fatty acids, and amino acids. Although its primary function is energy production, it also provides intermediates for synthesizing the amino acids and porphyrins.

Erythrocytes differ from other human cells in that they lack mitochondria, and therefore the TCA cycle is not found in them.

The TCA cycle oxidizes acetic acid completely to two moles CO_2 plus eight hydrogen atoms, which enter the electron-transport chain:

$$CH_3COOH + 2H_2O \longrightarrow 2CO_2 + 8H$$

Several sources provide the acetate to fuel the TCA cycle. The main source is the mitochondrial *pyruvate-dehydrogenase reaction*. Pyruvate formed in glycolysis or in the catabolism of five of the amino acids readily penetrates into the mitochondria. Thiamine pyrophosphate (TPP) decarboxylates pyruvate to CH_3CHOH-TPP. Oxidized lipoic acid, a second coenzyme, oxidizes this CH_3CHOH- to $CH_3C(=O)-$ and transfers it to CoA-SH, yielding acetyl-CoA and reduced lipoic acid. FAD returns lipoic acid to its oxidized state, generating $FADH_2$, which in turn reduces NAD^+ to NADH. Thus, five coenzymes participate in the pyruvate-dehydrogenase reaction. The overall reaction is:

$$\text{Pyruvate} + \text{CoA-SH} + NAD^+ \longrightarrow \text{acetyl-CoA} + CO_2 + NADH + H^+$$

ATP inhibits pyruvate dehydrogenase by phosphorylating the enzyme to render it inactive.

β-Oxidation of fatty acid, occurring within the mitochondria, is a second source of acetyl-CoA for the TCA cycle, as is the degradation of ketone bodies. Another source of acetyl-CoA is from the catabolism of eight of the amino acids.

As shown below and in Figure 11-1, *citrate synthase* brings acetate from acetyl-CoA into the TCA cycle by joining its methyl group to the keto group of

Figure 11-1

The tricarboxylic acid (TCA) cycle.

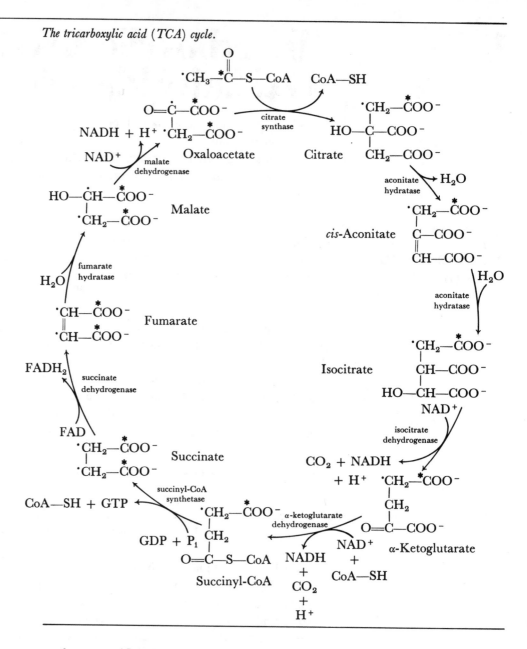

oxaloacetate (OAA) to produce citrate, a tricarboxylic acid. In Figure 11-1 and in the reactions to follow, the asterisk and dot markers are used to trace the fate of the carbon atoms of the carboxyl and methyl groups of the initial acetate:

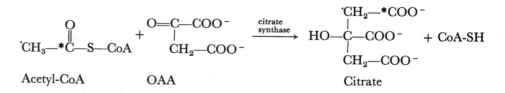

Because citrate synthase catalyzes the first committed step of the TCA cycle, it is not surprising that it controls the rate of the cycle. NADH and ATP allosterically inhibit citrate synthase.

Aconitate hydratase dehydrates citrate to create a carbon-carbon double bond in *cis*-aconitate (a hydratase adds water to a double bond or removes water to create a double bond). This hydratase then rehydrates *cis*-aconitate to yield isocitrate. The overall effect of these two reactions is to move the hydroxyl group of citrate to another carbon atom to create isocitrate.

Although citrate is symmetrical, aconitate hydratase can distinguish between the two ends of the molecule, because the enzyme binds to all three carboxyl groups. Hence, if the carbons of acetate are radioactively labeled, the labels will remain fixed in relative position, rather than appearing at both ends of the isocitrate molecules.

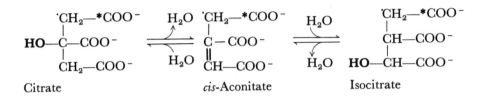

Citrate   *cis*-Aconitate   Isocitrate

Isocitrate dehydrogenase oxidizes the hydroxyl group of isocitrate to an α-keto group and removes the middle carboxyl group, yielding α-ketoglutarate (a five-carbon dicarboxylic acid), CO_2, and NADH. This NAD^+-linked isocitrate-dehydrogenase reaction is a rate-controlling step in the TCA cycle. Its positive allosteric modifier is ADP, while its negative modifiers are NADH and ATP. An $NADP^+$-linked isocitrate dehydrogenase also exists to create NADPH for fatty-acid and steroid biosynthesis. This enzyme does not participate in the TCA cycle.

$$
\begin{array}{c}
\overset{\cdot}{C}H_2\text{—}*COO^- \\
| \\
CH\text{—}COO^- \\
| \\
HO\text{—}CH\text{—}COO^-
\end{array}
+ NAD^+
\xrightarrow[\text{dehydrogenase}]{\text{isocitrate}}
\begin{array}{c}
^{\gamma}\overset{\cdot}{C}H_2\text{—}*COO^- \\
| \\
^{\beta}CH_2 \\
| \\
O\overset{\alpha}{=}C\text{—}COO^-
\end{array}
+ CO_2 + NADH + H^+
$$

Isocitrate α-Ketoglutarate

In the liver, α-ketoglutarate is transaminated to produce glutamate; conversely, the α-amino group is removed from glutamate to yield α-ketoglutarate. Thus, the pool of α-ketoglutarate fluctuates.

The *α-ketoglutarate-dehydrogenase* complex resembles that of pyruvate dehydrogenase; both decarboxylate and dehydrogenate a CoA-bound keto acid and produce NADH. α-Ketoglutarate dehydrogenase removes the α-carboxyl group of α-ketoglutarate to yield, in the presence of CoA-SH, succinyl-CoA:

$$
\begin{array}{c}
\overset{\cdot}{C}H_2\text{—}*COO^- \\
| \\
CH_2 \\
| \\
O\text{=}\overset{}{C}\text{—}\mathbf{COO^-}
\end{array}
+ CoA\text{-}SH + NAD^+ \rightleftharpoons
\begin{array}{c}
\overset{\cdot}{C}H_2\text{—}*COO^- \\
| \\
CH_2 \\
| \\
O\text{=}C\text{—}S\text{—}CoA
\end{array}
+ \mathbf{CO_2} + NADH + H^+
$$

α-Ketoglutarate   Succinyl-CoA

In the liver, succinyl-CoA is siphoned from the TCA cycle for porphyrin synthesis, but replaced by means of the degradation of certain amino acids to succinyl-CoA.

Succinyl-CoA synthetase removes CoA from succinyl-CoA to liberate succinate. This enzyme is named for the reverse reaction, the synthesis of succinyl-CoA. This cleavage of CoA drives the substrate-level phosphorylation of GDP to GTP; it is the only substrate-level phosphorylation in the TCA cycle. *Substrate-level phosphorylations* such as those in glycolysis, e.g., the conversion of PEP + ADP to pyruvate + ATP, were discussed in Chapter 10. Oxidative phosphorylation, in contrast, occurs by means of electron transport through the cytochrome system.

$$\begin{array}{c} \overset{\cdot}{C}H_2 - *COO^- \\ | \\ CH_2 \\ | \\ O = \overset{\cdot}{C} - S - CoA \end{array} + GDP + P_i \rightleftharpoons \begin{array}{c} \overset{\cdot}{C}H_2 - *COO^- \\ | \\ \overset{\cdot}{C}H_2 - *COO^- \end{array} + GTP$$

Succinyl-CoA   Succinate

The symmetry of succinate disperses the labels from the initial acetate molecule to both ends of the succinate molecules. Thus, both carboxyl groups of succinate are marked as being derived from the carboxyl group of acetate. Similarly, both —CH_2— groups are tagged as coming from the methyl group of acetate. The two CO_2 molecules liberated on the first turn of the cycle do not come from the initial acetate, but subsequent turns will liberate labeled CO_2 from the initial labeled acetate.

The GTP produced in the succinyl-CoA-synthetase reaction then phosphorylates ADP to ATP:

$$GTP + ADP \rightleftharpoons GDP + ATP$$

Succinate dehydrogenase oxidizes succinate with FAD to create the —CH=CH— bond of fumarate. Succinate dehydrogenase is the only FAD-linked dehydrogenase in the TCA cycle; isocitrate dehydrogenase, α-ketoglutarate dehydrogenase, and malate dehydrogenase are NAD^+-linked.

$$\begin{array}{c} \overset{\cdot}{C}H_2 - *COO^- \\ | \\ \overset{\cdot}{C}H_2 - *COO^- \end{array} + FAD \xrightarrow[\text{dehydrogenase}]{\text{succinate}} \begin{array}{c} \overset{\cdot}{C}H - *COO^- \\ \| \\ \overset{\cdot}{C}H - *COO^- \end{array} + FADH_2$$

Succinate   Fumarate

Fumarate hydratase then hydrates the —HC=CH— bond of fumarate to create malate:

$$\begin{array}{c} \overset{\cdot}{C}H - *COO^- \\ \| \\ \overset{\cdot}{C}H - *COO^- \end{array} + H_2O \xrightarrow[\text{hydratase}]{\text{fumarate}} \begin{array}{c} HO - \overset{\cdot}{C}H - *COO^- \\ | \\ \overset{\cdot}{C}H_2 - *COO^- \end{array}$$

Fumarate   Malate

To complete the cycle, *malate dehydrogenase* uses NAD^+ to oxidize the hydroxyl group of malate to the keto group of oxaloacetate:

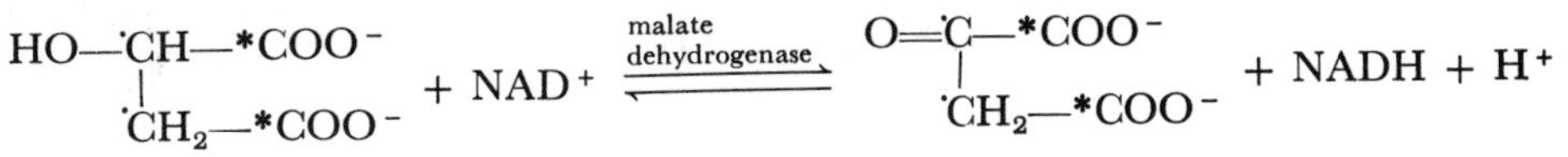

$$HO-CH-*COO^- \quad + NAD^+ \xrightarrow[\text{dehydrogenase}]{\text{malate}} \quad O=C-*COO^- \quad + NADH + H^+$$
$$\begin{array}{cc} | & \\ CH_2-*COO^- & \end{array} \qquad\qquad\qquad \begin{array}{cc} | & \\ CH_2-*COO^- & \end{array}$$

Malate   OAA

The TCA cycle yields three moles NADH, one mole FADH$_2$, and one mole GTP for every acetate unit it oxidizes.

TCA intermediates such as α-ketoglutarate and OAA are often removed from the cycle to serve as precursors for glutamate and aspartate, respectively. To replenish these intermediates, pyruvate carboxylase synthesizes OAA from pyruvate. In carbohydrate deficiency, there is not enough pyruvate for the pyruvate-carboxylase reaction, which causes depletion of the TCA-cycle intermediates and excess fat mobilization.

Problems 1–7

Match the structures below to their descriptions given in Problems 1–7:

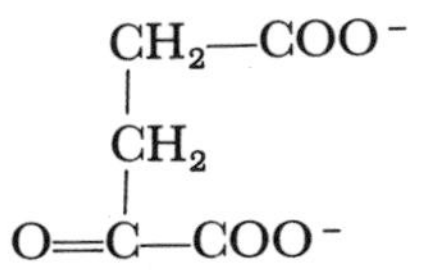

$$\begin{array}{c} CH_2-COO^- \\ | \\ CH_2 \\ | \\ O=C-COO^- \end{array} \qquad \begin{array}{c} O=C-COO^- \\ | \\ CH_2-COO^- \end{array} \qquad \begin{array}{c} CH_2-COO^- \\ | \\ HO-C-COO^- \\ | \\ CH_2-COO^- \end{array}$$

A. 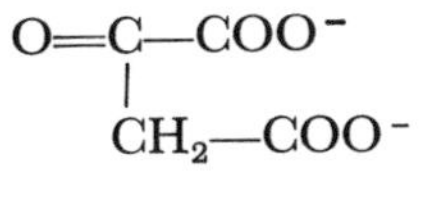   B. 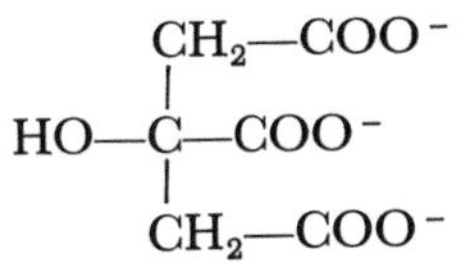   C.

$$\begin{array}{c} CH_2-COO^- \\ | \\ CH_2-COO^- \end{array}$$

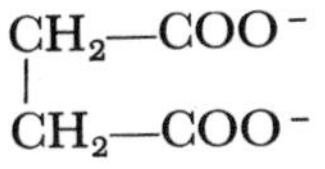

D.

1. Tricarboxylic acid(s)
2. α-Keto acid(s)
3. Dicarboxylic acid(s)
4. Oxaloacetate
5. Citrate
6. α-Ketoglutarate
7. Succinate

Problems 8–10

Predict the product(s) of each of the reactions in Problems 8–10:

8. Succinyl-CoA + GDP + P_i $\xrightarrow{\text{succinyl-CoA synthetase}}$ . . .

9. Isocitrate + NAD^+ $\xrightarrow{\text{isocitrate dehydrogenase}}$ . . .

10. Succinate + FAD $\xrightarrow{\text{succinate dehydrogenase}}$ . . .

Name the structures and the enzymes for the reactions in Problems 11–13:

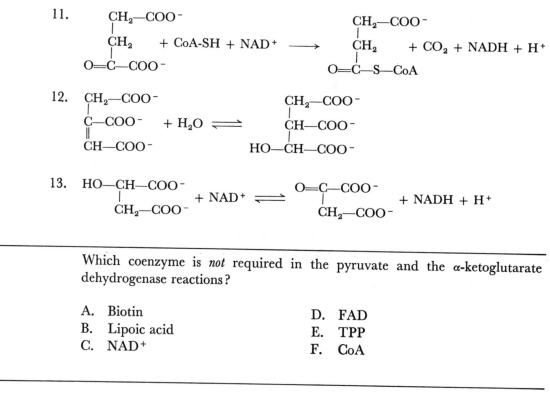

11.

$$CH_2-COO^- \quad CH_2-COO^-$$
$$CH_2 \quad + CoA\text{-}SH + NAD^+ \longrightarrow CH_2 \quad + CO_2 + NADH + H^+$$
$$O{=}C{-}COO^- \qquad O{=}C{-}S{-}CoA$$

12.

$$CH_2-COO^- \qquad CH_2-COO^-$$
$$C{-}COO^- \quad + H_2O \rightleftharpoons CH{-}COO^-$$
$$CH{-}COO^- \qquad HO{-}CH{-}COO^-$$

13.

$$HO{-}CH{-}COO^- \quad O{=}C{-}COO^-$$
$$\qquad\qquad\qquad + NAD^+ \rightleftharpoons \qquad\qquad + NADH + H^+$$
$$CH_2{-}COO^- \qquad CH_2{-}COO^-$$

Problem 14

Which coenzyme is *not* required in the pyruvate and the α-ketoglutarate dehydrogenase reactions?

A. Biotin
B. Lipoic acid
C. NAD^+

D. FAD
E. TPP
F. CoA

OXIDATIVE PHOSPHORYLATION

Porphyrins and the Heme-Proteins

The players in the oxidative-phosphorylation game consist of heme-proteins as well as non-heme, iron-containing proteins.

The fundamental unit in the structure of heme is the pyrrole ring, shown below. Four pyrrole rings are linked to one another by $={=}$CH$—$ bridges to create porphin (shown below), the parent compound to the porphyrins. Side chains are added to the porphin ring to yield protoporphyrin. Ferrous iron (Fe^{+2}) then binds to the central four nitrogen atoms of protoporphyrin to yield heme:

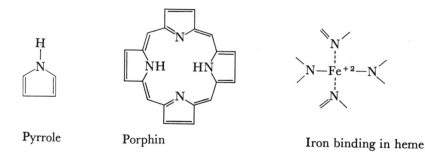

Pyrrole   Porphin   Iron binding in heme

Heme, when conjugated with a protein, forms a *heme-protein*, or hemoprotein; the major heme-proteins are hemoglobin, an O_2 carrier, and the cytochromes, the electron carriers. Cytochrome P_{450}, which is found in liver microsomes, differs from other cytochromes in that it does not participate in oxidative phosphorylation.

There are two hydrogen-electron carriers in oxidative phosphorylation that contain non-heme, iron proteins. Their iron atom binds directly to the enzyme without requiring a porphyrin ring. These carriers are succinate dehydrogenase and NADH dehydrogenase.

Humans synthesize porphyrins and heme from glycine and the TCA intermediate, succinyl-CoA. Aminolevulinic acid (ALA) synthetase uses pyridoxal phosphate as a coenzyme to join glycine with succinyl-CoA to yield ALA. Vitamin B_6 deficiency produces hypochromic microcytic anemia, which is the result of inadequate hemoglobin synthesis, by depriving ALA synthetase of its coenzyme.

The ALA synthetase reaction controls the rate of porphyrin synthesis. Heme, the endproduct of porphyrin synthesis, inhibits ALA synthetase. ALA is converted to porphobilinogen, a compound consisting of a pyrrole nucleus with attached side-chains. Four molecules of porphobilinogen then join to form an uroporphyrin. Uroporphyrins are decarboxylated to coproporphyrins, which are then oxidized to protoporphyrins. Heme synthetase adds Fe^{+2} to protoporphyrin to yield heme.

The *porphyrias* are a class of acquired or inherited disorders characterized by increased serum levels and urinary excretion of porphyrins. When exposed to sunlight, urine specimens from patients with such disorders will often darken.

Lead poisoning, which includes porphyria among its symptoms, inhibits several enzymes of porphyrin synthesis, particularly heme synthetase. The inhibition of heme synthetase results in hypochromic microcytic anemia and the overproduction of coproporphyrins and ALA. Attacks of acute abdominal pain and neurologic impairment also ensue.

The inherited porphyrias usually stem from an overactive ALA synthetase enzyme that is not inhibited by heme. Except for acute intermittent porphyria, the inherited porphyrias cause photosensitivity dermatitis, possibly due to fluorescence of the excess porphyrins. Acute intermittent porphyria causes attacks of abdominal pain similar to those in lead poisoning.

Heme Degradation

In degrading each hemoglobin molecule, the reticuloendothelial cells of the liver, spleen, and bone marrow separate the four heme molecules from the two α and two β polypeptide chains of the protein part of the molecule. The iron from heme is removed to create biliverdin, which is reduced to bilirubin. Of the heme catabolized to bilirubin, the vast majority arises from hemoglobin degradation. Lesser amounts are derived from cytochrome breakdown.

Albumin binds to bilirubin and transports it to the liver. There, glucuronyl transferase conjugates two moles of UDP-glucuronic acid to each bilirubin molecule to create bilirubin diglucuronide, or conjugated bilirubin.

Premature infants frequently develop jaundice (defined as hyperbilirubinemia), because their liver has not yet produced adequate amounts of glucuronyl transferase.

The indirect van den Bergh test measures free or unconjugated bilirubin, whereas the direct van den Bergh test assays conjugated bilirubin.

The liver excretes bilirubin diglucuronide into the bile. When it reaches the intestines, bacteria reduce it to urobilinogen. Intestinal urobilinogen has one of two fates. Some is absorbed into the plasma and may be either reexcreted into the bile or excreted into the urine, where it is oxidized to urobilin. The remainder is reduced to stercobilinogen and stercobilin in the feces.

Oxidative Phosphorylation and the Electron-Transport Chain

Oxidative phosphorylation involves the production of ATP from ADP and P_i by harnessing the energy released as electrons are transferred during a series of oxidation-reduction reactions. Because $\Delta E^{0'}$ for each electron transfer is positive, ΔG is negative (as shown by the formula below) and the reactions will occur spontaneously under standard conditions:

$$\Delta G^{0'} \ (\text{kcal/mole}) = -23.1n\Delta E^{0'}$$

Like the TCA cycle, oxidative phosphorylation is lodged exclusively within the mitochondria.

The initial carriers in electron transport—NADH dehydrogenase and succinate dehydrogenase—are flavin-linked and contain non-heme, iron proteins. As shown in Figure 11-2 (pathway *A*), the succinate dehydrogenase of the TCA cycle oxidizes succinate to fumarate, thereby reducing FAD to $FADH_2$, which in turn reduces CoQ (ubiquinone) to $CoQH_2$ (ubiquinol).

Problem 15

Malonate is a three-carbon dicarboxylic acid that inhibits succinate dehydrogenase. Is it a competitive, noncompetitive, or irreversible inhibitor of this enzyme?

In pathway *B* of Figure 11-2, NADH dehydrogenase oxidizes NADH to NAD^+, thereby reducing FMN to $FMNH_2$. The flavoprotein formed by conjugation of $FMNH_2$ to the enzyme then transfers the two hydrogen atoms from its riboflavin component to CoQ.

Pathways *C* and *D* (Fig. 11-2) show other routes for reducing CoQ. These are the β-oxidation of fatty acids and the mitochondrial oxidation of glycerol-3-P to DHAP. The NADH-dehydrogenase pathway is the only one that produces ATP before transferring its hydrogens and electrons to CoQ.

Pathway *E* shows that $CoQH_2$ transfers an electron pair to two ferric (Fe^{+3}) ions of two cytochrome b molecules, reducing them to ferrous (Fe^{+2}) ions. As cytochrome b transfers its electrons to the Fe^{+3} of cytochrome c_1, ADP is phosphorylated with P_i to create ATP. Antimycin A, an antibiotic (though never used internally) and fungicide, stops electron transport from cytochrome b to cytochrome c_1, thereby halting the process of oxidative phosphorylation.

Electron transfer and oxidative phosphorylation by NADH dehydrogenase continues until all the CoQ and cytochrome b are reduced. The last step in oxidative phosphorylation occurs when cytochrome a reduces cytochrome a_3. Cytochrome a_3 then transfers its electron pair to $\frac{1}{2}O_2$ and $2H^+$ in the medium to create H_2O.

Together, cytochromes a_3 and a are called *cytochrome oxidase*. Cytochrome oxidase contains copper in addition to iron. Cyanide is the most common poison that blocks electron transport at the cytochrome oxidase step.

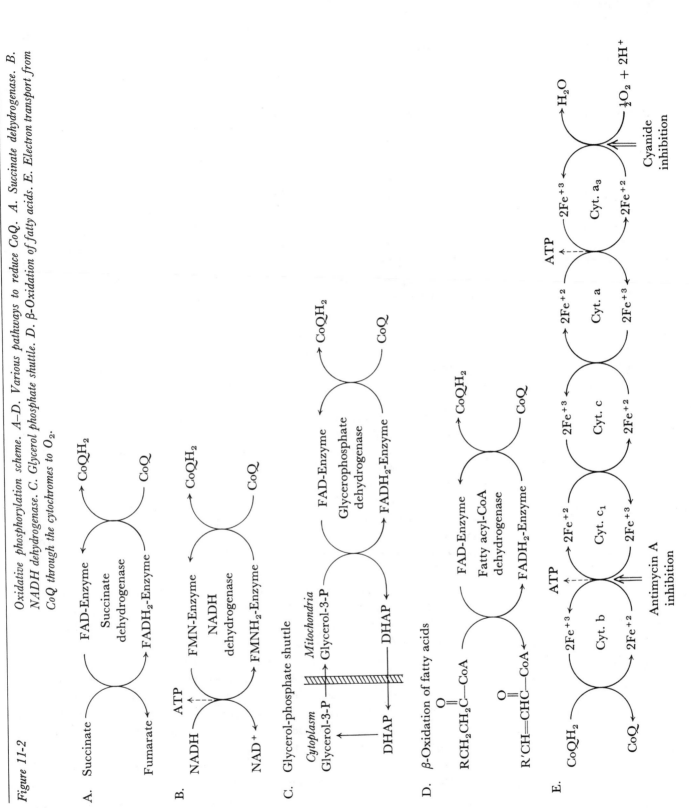

Figure 11-2   Oxidative phosphorylation scheme. A–D. Various pathways to reduce CoQ. A. Succinate dehydrogenase. B. NADH dehydrogenase. C. Glycerol phosphate shuttle. D. β-Oxidation of fatty acids. E. Electron transport from CoQ through the cytochromes to O_2.

When NADH is oxidized by NADH dehydrogenase in electron transport, three moles of ATP are produced, and the net oxidation-reduction reaction is:

$$
\begin{array}{ll}
NADH \longrightarrow NAD^+ + H^+ + 2e^- & E = -E^{0\prime} = +0.320 \text{ V} \\
\tfrac{1}{2}O_2 + 2H^+ + 2e^- \longrightarrow H_2O & E = E^{0\prime} = +0.816 \text{ V} \\
\hline
Net:\ NADH + \tfrac{1}{2}O_2 + H^+ \longrightarrow NAD^+ + H_2O & \Delta E^{0\prime} = +1.14 \text{ V}
\end{array}
$$

$$\Delta G^{0\prime} = -23.1(n)(\Delta E^{0\prime}) = -23.1(2)(1.14) = -52.7 \text{ kcal/mole}$$

To phosphorylate three moles ADP to three moles ATP, $+21.9$ kcal must be consumed (3×7.3 kcal/mole). Hence, the efficiency of electron transport is 21.9 kcal/52.7 kcal per mole NADH, or about 40%.

Problem 16

When succinate is oxidized in the electron-transport chain, only two moles of ATP are produced, compared to three moles ATP produced per NADH oxidized. Explain this difference by calculating $\Delta E^{0\prime}$ and $\Delta G^{0\prime}$ for the succinate dehydrogenase reaction:

$$
\begin{array}{ll}
Succinate \longrightarrow fumarate + 2H^+ + 2e^- & E = -E^{0\prime} = +0.031 \text{ V} \\
\tfrac{1}{2}O_2 + 2H^+ + 2e^- \longrightarrow H_2O & E = E^{0\prime} = +0.816 \text{ V} \\
\hline
Net:\ Succinate + \tfrac{1}{2}O_2 \longrightarrow fumarate + H_2O &
\end{array}
$$

Since it is unable to penetrate the mitochondria itself, cytoplasmic NADH utilizes several shuttles to transfer its electrons into the mitochondria. The *glycerol-phosphate shuttle*, shown as pathway C in Figure 11-2, uses glycerol-3-P to convey an electron pair from cytoplasmic NADH into the mitochondria. Glycerol-3-P readily penetrates the outer mitochondrial membrane to reach glycerol-phosphate dehydrogenase on the inner membrane, where it is re-oxidized to DHAP, thereby reducing FAD to FADH$_2$. For every mole of cytoplasmic NADH so oxidized, only two moles ATP are generated in electron transport, compared to the three moles ATP produced per NADH in the purely mitochondrial reaction.

The *malate shuttle* employs malate to carry electrons into the mitochondria. Once inside, malate reduces NAD$^+$ to NADH as the malate is oxidized to OAA. This NADH, in turn, is oxidized by NADH dehydrogenase to yield three moles of ATP per mole of cytoplasmic NADH.

Since no one is certain which shuttle predominates, each cytoplasmic NADH that is produced from glycolysis may yield either two or three moles ATP. Thus, the net energy gain in glycolysis to pyruvate is worth either six or eight moles ATP per mole of glucose. The pyruvate-dehydrogenase reaction generates two moles NADH per mole of glucose, worth six moles ATP. The TCA cycle receives two acetate groups from every molecule of glucose converted to acetyl-CoA. Two turns of the TCA cycle, therefore, produce six moles NADH (worth 18 moles ATP), two moles FADH$_2$ (worth four moles ATP), and two moles GTP (worth two moles ATP); a total energy gain equivalent to 24 moles of ATP results from oxidizing these two acetate groups. Therefore, *the complete oxidation of glucose to CO_2 and H_2O yields either 36 or 38 moles ATP.* This is to be compared with the meager two moles of ATP produced in anaerobic glycolysis, when the TCA cycle and oxidative phosphorylation cannot operate. Thus, exercising muscle, being anaerobic in its metabolism, must convert glucose to lactate at a terrific pace to compensate for this low energy yield.

The rate-controlling factor in electron transport is the availability of ADP. Under normal circumstances, O_2, NADH, $FADH_2$, and P_i abound and do not influence the rate of electron transport. Electron transport is tightly coupled to oxidative phosphorylation; when ADP is scarce, electron transport ceases. This coupling is called *acceptor regulation* or *respiratory control.*

When electron transport proceeds without concomitant ATP production, the reactions of the mitochondria are said to be *uncoupled.* Such a loss of acceptor control occurs either after exposure to uncoupling agents, such as 2,4-dinitrophenol (DNP), or after mechanical damage. Mitochondria extracted from cells by centrifugation exhibit a progressive loss of acceptor control, and within twelve hours, their reactions become uncoupled. The energy released in electron transport by uncoupled mitochondrial reactions is liberated as heat.

Aspirin overdose partially uncouples oxidative phosphorylation, and it thus raises the body temperature as a result of consequent heat production. An excess of thyroid hormone also partially uncouples oxidative phosphorylation.

Inhibitors of electron transport, such as cyanide or antimycin A, *do not* uncouple. Instead, they block electron transport, thereby stopping oxidative phosphorylation.

The *P : O ratio* for a substrate is defined as the ratio of P_i consumed in ATP formation per oxygen atom consumed. The P:O ratio for NADH is three, while for $FADH_2$, it is two. Uncoupled mitochondria have a P:O ratio of zero for any substrate.

Problem 17

Severe thiamine deficiency, known as "wet" beriberi, impairs myocardial metabolism and causes congestive heart failure by removing an essential coenzyme from which of the following reactions:

A. the electron-transport chain
B. the pyruvate and α-ketoglutarate dehydrogenase reactions
C. the isocitrate-dehydrogenase reaction
D. the glutamate-transaminase reaction

Problem 18

Which electron transfer is *not* accompanied by oxidative phosphorylation?

A. Cytochrome b → cytochrome c_1
B. NADH dehydrogenase → CoQ
C. Succinate dehydrogenase → CoQ
D. Cytochrome a → cytochrome a_3

Problems 19–21

Match the compounds below to their descriptions in Problems 19–21. You may choose more than one answer for each statement.

A. Aspirin overdose
B. Cyanide
C. 2,4-Dinitrophenol
D. Antimycin A
E. Thyroid hormones at toxic levels

19. Inhibits electron transfer from cytochrome b to cytochrome c_1

20. Uncouples oxidative phosphorylation

21. Inhibits cytochrome oxidase

Problem 22

Account for each of the 36 to 38 moles of ATP produced in the complete oxidation of one mole of glucose to CO_2 plus H_2O.

Problem 23

When radioactively labeled $^{14}CH_3C(=O)$-CoA enters the TCA cycle, where will the label appear in α-ketoglutarate during the first revolution of the cycle?

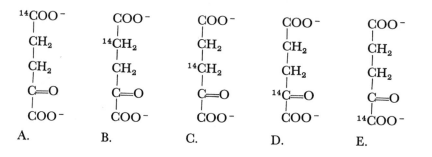

Problem 24

When $CH_3—^{14}C(=O)$-CoA is fed into the TCA cycle, where will this label appear in malate during the first turn of the cycle?

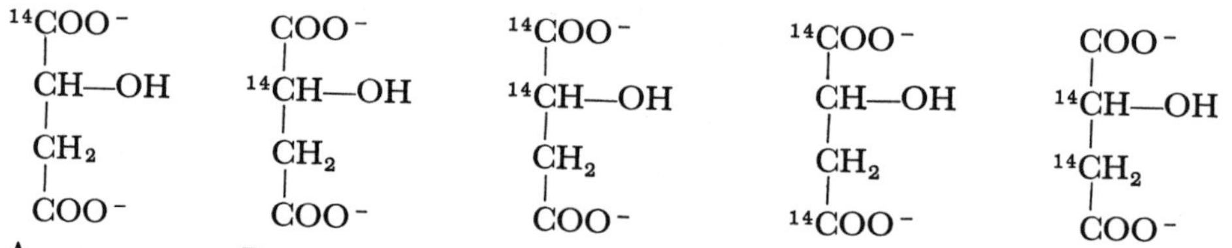

$^{14}COO^-$	COO^-	$^{14}COO^-$	$^{14}COO^-$	COO^-
$CH—OH$	$^{14}CH—OH$	$^{14}CH—OH$	$CH—OH$	$^{14}CH—OH$
CH_2	CH_2	CH_2	CH_2	$^{14}CH_2$
COO^-	COO^-	COO^-	$^{14}COO^-$	COO^-
A.	B.	C.	D.	E.

Problem 25

The enzymes of the two major rate-controlling steps of the TCA cycle are:

A. Malate dehydrogenase
B. α-Ketoglutarate dehydrogenase
C. Isocitrate dehydrogenase
D. Succinate dehydrogenase
E. Citrate synthase

1. C.
2. A, B.
3. A, B, D.
4. B. (This molecule may be thought of as oxalate, $^-OOC—COO^-$, linked to acetate, $CH_3—COO^-$.
5. C.
6. A.
7. D.
8. Succinate + CoA-SH + GTP
9. α-Ketoglutarate + CO_2 + NADH + H^+
10. Fumarate + $FADH_2$
11. The left-hand structure is α-ketoglutarate. On the right is succinyl-CoA. Since this reaction involves dehydrogenation and decarboxylation, the enzyme complex might be named α-ketoglutarate dehydrogenase or decarboxylase. By convention, it is called a dehydrogenase. Five coenzymes participate: CoA, NAD^+, FAD, TPP, and lipoic acid.
12. On the left is *cis*-aconitate; on the right is isocitrate. Aconitate hydratase catalyzes this hydration reaction.
13. On the left is malate, while on the right is oxaloacetate. Malate dehydrogenase catalyzes this oxidation.
14. A. (Biotin is used in carboxylation reactions, whereas both of these reactions are decarboxylations.)
15. Competitive.

16. $\Delta E^{0'} = +0.847$ V

    $\Delta G^{0'} = -23.1(n)(\Delta E^{0'})$

    $\qquad = -23.1(2)(+0.847) = -39.2$ kcal/mole

    These values of $\Delta E^{0'}$ and $\Delta G^{0'}$ are only about 75% as great as those for NADH oxidation in electron transport. Assuming 50% efficiency in utilizing the free energy released, only 19.6 kcal of work could be performed per mole of succinate oxidized to fumarate—enough to phosphorylate two moles ADP (14.6 kcal) but too little to drive the synthesis of three moles ATP (21.9 kcal).

17. B. (Lack of TPP slows the TCA cycle and forces the heart to rely on anaerobic glycolysis with its low-energy yield.)
18. C. (The succinate-dehydrogenase reaction utilizes FAD and is not accompanied by the production of ATP.)
19. D.
20. A, C, E.
21. B.
22. See the section entitled "Energetics of Oxidative Phosphorylation."
23. B.
24. D.
25. C, E.

REFERENCES

Lehninger, A. L. *Biochemistry: The Molecular Basis of Cell Structure and Function* (2nd ed.). New York: Worth, 1975. Pp. 433–465, 477–524, 533–537.

Marver, H., and Schmid, R. The Porphyrias. In Stanbury, J., Wyngaarden, J., and Fredrickson, D. (Eds.), *The Metabolic Basis of Inherited Disease* (3rd ed.). New York: McGraw-Hill, 1972.

Mathews-Roth, M. M. Erythropoietic protoporphyria—Diagnosis and Treatment. *N. Eng. J. Med.* 297:98, 1977.

Schmid, R. Hyperbilirubinemia. In Stanbury, J., Wyngaarden, J., and Fredrickson, D. (Eds.), *The Metabolic Basis of Inherited Disease* (3rd ed.). New York: McGraw-Hill, 1972.

White, A., Handler, P., and Smith, E. L. *Principles of Biochemistry* (5th ed.). New York: McGraw-Hill, 1973. Pp. 166–179, 337–367, 664–669.

12 Lipid Metabolism and Biosynthesis

Atherosclerosis, the deposition of lipid plaques on the lining of arteries, is the leading cause of death in America. Because of the association between atherosclerosis and hyperlipoproteinemia (elevated serum lipoproteins), an extensive campaign of medical research has been launched to explore lipid metabolism.

This chapter deals with the metabolism and biosynthesis of six general types of lipids: triglycerides, fatty acids, ketone bodies, cholesterol, phosphoglycerides, and sphingolipids.

LIPOLYSIS

Lipolysis, defined as triglyceride hydrolysis, liberates fatty acids from their main storage depots in the triglycerides. Lipolysis begins with the intestinal hydrolysis of dietary triglycerides by pancreatic lipase. Once absorbed into the intestinal mucosa, the resultant free fatty acids (FFA), glycerol, and monoglycerides are resynthesized into triglycerides, which combine with lesser amounts of protein, phospholipid, and cholesterol to create chylomicrons. Plasma lipoprotein lipase hydrolyzes triglycerides in the chylomicrons into FFA and glycerol. Adipose tissue contains a hormone-sensitive lipase that hydrolyzes its triglycerides.

The glycerol released in lipolysis is phosphorylated to glycerol-3-P. For every two moles of glycerol-3-P, one mole of glucose can be synthesized in gluconeogenesis. Glycerol-3-P also serves as a precursor to triglyceride synthesis.

TRIGLYCERIDE SYNTHESIS

The intestinal mucosa differs from other sites of triglyceride synthesis in having a large supply of monoglycerides; each mole of monoglyceride may be combined with two moles of fatty-acyl-CoA to produce the corresponding triglyceride. Acyl-CoA synthetase activates fatty acids by linking them to the SH group of CoA-SH to create fatty-acyl-CoA in the cytoplasm:

$$R—COO^- + CoA\text{-}SH + ATP \rightleftharpoons R—\overset{\overset{\displaystyle O}{\|}}{C}—S—CoA + AMP + PP_i$$

In the liver and adipose tissue, triglycerides are synthesized from phosphatidic acids. Two pathways produce phosphatidic acids. In the first, two moles of fatty-acyl-CoA add to glycerol-3-P to yield the phosphatidic acid. The second route links two moles of fatty-acyl-CoA to DHAP by means of reducing its keto group with NADPH, thus forming phosphatidic acid, as shown in Figure 12-1. After synthesis via either pathway, phosphate is then cleaved from the phosphatidic acid, leaving a diglyceride. A third fatty-acyl-CoA is esterified to this diglyceride to yield a triglyceride.

Clofibrate is a drug that is used clinically to lower the plasma triglyceride level in hypertriglyceridemia. Although its mechanism of action is uncertain, it seems to augment the rate of triglyceride removal from the blood.

Figure 12-1

Triglyceride synthesis in liver and adipose tissue.

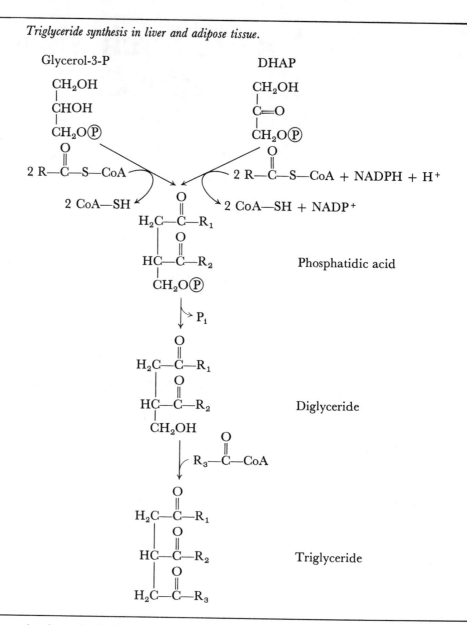

Phosphatidic acid

Diglyceride

Triglyceride

BETA-OXIDATION OF FATTY ACIDS

As the principal route for catabolizing fatty acids, β-oxidation occurs in the mitochondria, the intracellular "powerhouses." *β-Oxidation* is so named because it oxidizes the β-carbon atom of a fatty acid to a β-keto acid.

α-Oxidation of fatty acids occurs in the human brain. The rare inherited absence of an enzyme required for α-oxidation causes Refsum's disease.

ω-Oxidation of fatty acids is a minor pathway that is found in the liver.

Fatty-acyl-CoA from the cytoplasm cannot penetrate the mitochondrial membranes, because these membranes are impermeable to CoA. Fatty acids enter the mitochondria after binding to carnitine to produce fatty-acyl-carnitine. Once inside, CoA-SH displaces carnitine to regenerate fatty-acyl-CoA.

β-Oxidation, shown in Figure 12-2, basically consists of four reactions that can be summarized as follows:

1. $-\overset{\beta}{C}H_2-\overset{\alpha}{C}H_2- + FAD \xrightarrow{\text{Dehydrogenation}} -CH=CH- + FADH_2$

Figure 12-2

β-*Oxidation of fatty acids.*

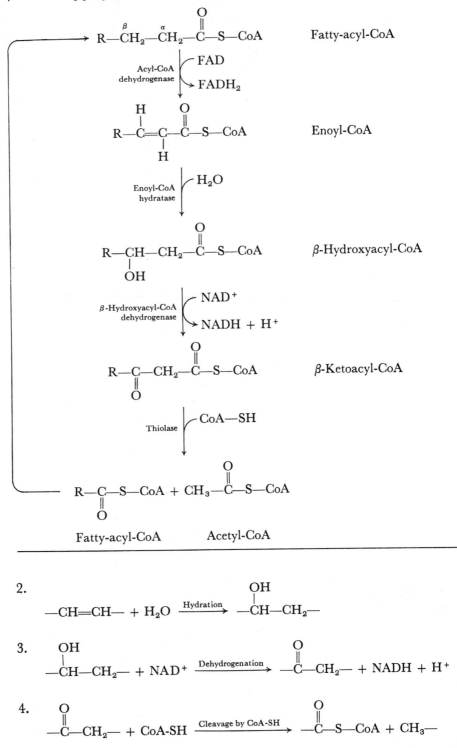

The pathway of β-oxidation begins as acyl-CoA dehydrogenase oxidizes the fatty acid to create a *trans* double bond between the α and β carbon atoms, thereby reducing FAD to $FADH_2$ (see Fig. 12-2). This monoenoic compound is named enoyl-CoA. A hydratase then hydrates this double bond, **yielding** β-hydroxyacyl-CoA. A second dehydrogenase oxidizes this β-hydroxy group to a β-keto group, creating β-ketoacyl-CoA and reducing NAD^+ to NADH (this step lends the name "β-oxidation" to the pathway). Finally, a thiolase uses the SH binding of CoA-SH to cleave the bond between the α and β carbon atoms, liberating acetyl-CoA. The remaining fatty-acyl-CoA, with two less carbon atoms than the original, can then reenter β-oxidation.

Each pass through β-oxidation removes two carbons from the fatty acid as acetyl-CoA and produces one mole $FADH_2$ and one mole NADH, which yield five moles ATP after reoxidation to FAD and NAD^+ in the electron-transport chain (two moles ATP per mole of $FADH_2$ oxidized and three moles ATP per mole of NADH oxidized). The complete β-oxidation of a 12-carbon, saturated fatty acid, for example, requires five passes through this sequence; hence, six moles acetyl-CoA, five moles $FADH_2$, and five moles NADH are produced. Five moles of O_2 are utilized by electron transport to oxidize these $FADH_2$ and NADH and yield 25 moles ATP. Since one mole of ATP was consumed to activate the fatty acid—i.e., to convert it to fatty-acyl-CoA—the net ATP yield is 24 moles. If the six moles of acetyl-CoA produced by β-oxidation are oxidized in the TCA cycle to 12 moles CO_2, then six moles GTP, six moles $FADH_2$, and 18 moles NADH are generated. Twelve moles of O_2 are utilized to oxidize these $FADH_2$ and NADH, and thus 72 moles ATP are generated from the total oxidation of these six acetyl-CoA molecules. Therefore, a total of 96 moles ATP and 12 moles CO_2 are produced, while 17 moles O_2 are consumed.

The *respiratory quotient (RQ)* for a catabolized substance is defined as the moles of CO_2 produced divided by the moles of O_2 consumed. For this hypothetical 12-carbon, saturated fatty acid, then, the RQ is:

$$RQ = \frac{12 \ CO_2 \ produced}{17 \ O_2 \ consumed} = 0.71$$

The complete oxidation of a 12-carbon disaccharide using glycolysis and the TCA cycle produces only 72 or 76 moles ATP, compared to the 96 moles ATP gained by oxidizing a 12-carbon fatty acid. For this disaccharide, since 12 moles CO_2 are produced while 12 moles O_2 are consumed, its RQ is 1.0.

Biologists calculate the RQ for animals to determine their chief metabolic fuel; an RQ near 1.0 denotes reliance on glycolysis, whereas an RQ near 0.7 indicates a reliance on β-oxidation of fatty acids. When the Caloric values of food are calculated, triglyceride is found to have 9 Cal/gram compared to only 4 Cal/gram for carbohydrate and protein. This explains why mammals rely on triglyceride rather than carbohydrate for energy storage.

CATABOLISM OF ODD-CARBON FATTY ACIDS

Most naturally occurring fatty acids have an even number of carbon atoms and are therefore completely degraded to acetyl-CoA during β-oxidation. However, propionate, a three-carbon fatty acid, arises from the β-oxidation of odd-carbon fatty acids and from the catabolism of two branched-chain amino acids, isoleucine and valine.

In propionate metabolism, a carboxylase adds CO_2 to propionyl-CoA to create methylmalonyl-CoA. A mutase then transfers the —C(=O)—S—CoA group to the methyl group, yielding succinyl-CoA:

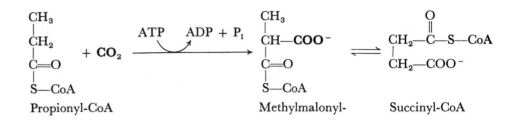

Propionyl-CoA   Methylmalonyl-   Succinyl-CoA

Problems 1–2

Choose the coenzymes required by the enzymes of propionate metabolism given below:

A. TPP
B. Deoxyadenosylcobalamin
C. Lipoic acid
D. Biotin

1. Propionyl-CoA carboxylase
2. Methylmalonyl-CoA mutase

Problem 3

The fates of succinyl-CoA in metabolism include two of the following:

A. β-Oxidation
B. Porphyrin biosynthesis
C. Fueling a substrate-level phosphorylation in the TCA cycle
D. Transamination to form glutamate

Problems 4–7

Draw the missing structure in each of the following reactions of β-oxidation, describe the reaction, and name the enzyme type (dehydrogenase, hydratase, or thiolase):

4.
$$R-\overset{O}{\overset{\|}{C}}-CH_2-\overset{O}{\overset{\|}{C}}-S-CoA + CoA\text{-}SH \longrightarrow \cdots$$

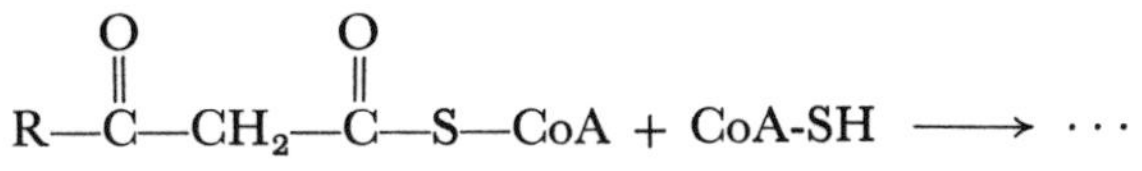

5.
$$\cdots \longrightarrow R-\overset{OH}{\overset{|}{C}H}-CH_2-\overset{O}{\overset{\|}{C}}-S-CoA$$

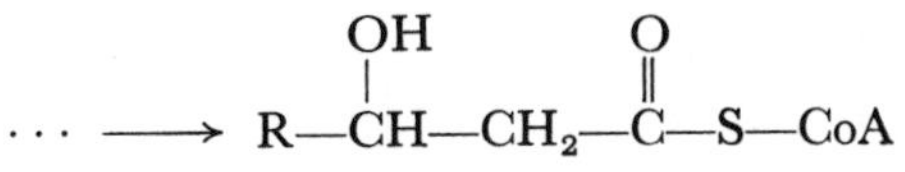

6.
$$R-CH_2-CH_2-\overset{O}{\overset{\|}{C}}-S-CoA + FAD \longrightarrow \cdots$$

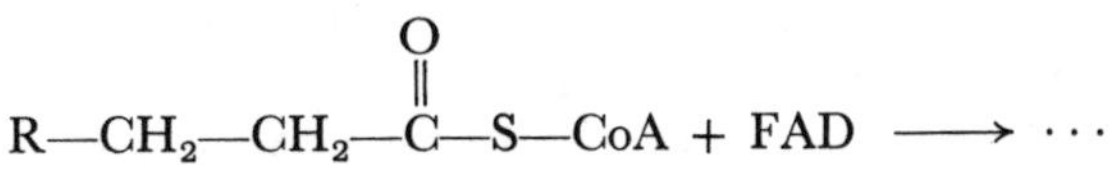

7.
$$R-\overset{OH}{\overset{|}{C}H}-CH_2-\overset{O}{\overset{\|}{C}}-S-CoA + NAD^+ \longrightarrow \cdots$$

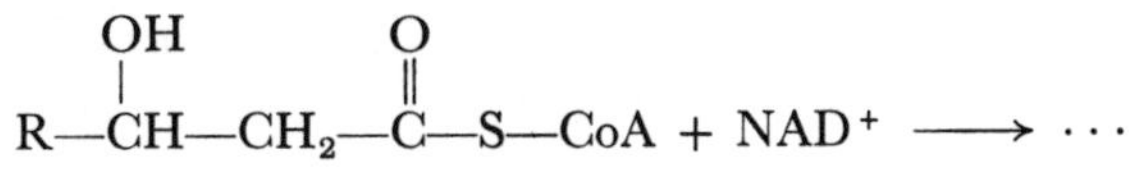

In the β-oxidation of myristic acid (14 carbons) to form acetyl-CoA:

A. How many moles of ATP are consumed to activate myristic acid by joining it to CoA?
B. How many passes through β-oxidation are required to yield seven moles acetyl-CoA?
C. How many moles of $FADH_2$ and NADH are produced?
D. What is the net ATP yield after the $FADH_2$ and NADH have been re-oxidized in electron transport, and how many moles O_2 are consumed?

If the seven moles acetyl-CoA from Problem 8 were oxidized to CO_2 + H_2O in the TCA cycle:

A. How many moles GTP would be produced?
B. How many moles of $FADH_2$ and NADH would be produced?
C. How many moles of ATP would be produced from the TCA cycle, and how many moles O_2 would be consumed?
D. For the oxidation of myristic acid to CO_2 + H_2O, how many net moles ATP are produced, and what is the RQ of this acid?

CYTOPLASMIC SYNTHESIS OF SATURATED FATTY ACIDS

In humans, saturated fatty acids are synthesized from bicarbonate and acetyl-CoA. This cytoplasmic biosynthesis requires NADPH, which is supplied chiefly by the hexose-monophosphate shunt. Acyl-carrier protein (ACP) is employed to carry the elongating fatty acid chain. Like CoA, ACP contains the B vitamin, pantothenic acid.

The first step toward this synthesis is the carboxylation of acetyl-CoA by acetyl-CoA carboxylase and biotin to produce malonyl-CoA. Malonate is a three-carbon, dicarboxylic acid.

$$CH_3 \overset{\overset{\text{O}}{\|}}{-C}-S-CoA + HCO_3^- + ATP \rightleftharpoons$$

$$^-OOC-CH_2 \overset{\overset{\text{O}}{\|}}{-C}-S-CoA + ADP + P_i$$

Acetyl-CoA

Malonyl-CoA

Acetyl-CoA carboxylase is an allosteric enzyme that controls the rate of fatty acid synthesis. It is logical that citrate and isocitrate are its positive modulators, because when these TCA-cycle intermediates abound, the cell is rich in energy and can afford to synthesize fatty acids. In addition, some of the citrate inside the mitochondria enters the cytoplasm, where citrate lyase degrades it to acetyl-CoA and OAA (the reverse of the citrate synthase re-action). This citrate-lyase reaction provides much of the acetyl-CoA to fuel fatty acid biosynthesis. In addition, carnitine transfers acetyl groups from the mitochondria to the cytoplasm for use in fatty acid biosynthesis.

Figure 12-3

Cytoplasmic fatty-acid biosynthesis.

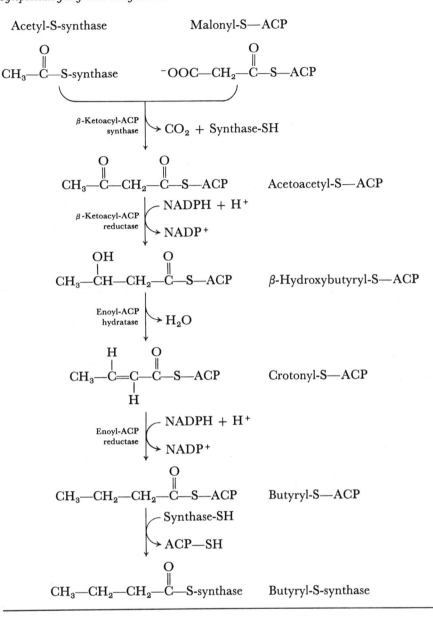

Acetate and malonate are transferred from CoA to ACP by the below reactions. Acetate is then transferred from ACP to a synthase enzyme.

Acetyl-S-CoA + ACP-SH $\rightleftharpoons$ acetyl-S-ACP + CoA-SH

Acetyl-S-ACP + synthase-SH $\rightleftharpoons$ acetyl-S-synthase + ACP-SH

Malonyl-S-CoA + ACP-SH $\rightleftharpoons$ malonyl-S-ACP + CoA-SH

As shown in Figure 12-3, acetyl-S-synthase initiates fatty acid biosynthesis by condensing with malonyl-S-ACP, generating acetoacetyl-S-ACP. In this condensation, malonate liberates its carboxyl group as CO_2; this is the same carbon atom that was added from HCO_3^- to synthesize malonyl-CoA.

The remaining reactions of fatty acid biosynthesis accomplish the reverse of the β-oxidation sequence, though using different enzymes. β-Ketoacyl reductase uses NADPH to reduce the β-keto group of acetoacetyl-S-ACP to a β-hydroxyl group, giving rise to β-hydroxylbutyryl-S-ACP. A hydratase removes this β-hydroxyl group along with a hydrogen atom, creating the carbon-carbon double bond of crotonyl-S-ACP. Enoyl reductase then saturates this double bond using NADPH to yield butyryl-S-ACP.

The cycle then repeats itself. Butyrate is transferred from ACP to the synthase enzyme and condenses with malonyl-S-ACP. Ultimately, a 16-carbon compound, palmityl-S-ACP, is generated, and a thioesterase cleaves this to palmitic acid plus ACP-SH.

For each turn of the cycle, two moles of NADPH are consumed by the two reductase reactions, and one mole of ATP is consumed in synthesizing malonyl-CoA. Since each mole of NADPH could yield three moles ATP in oxidative phosphorylation, each turn of the cycle consumes seven moles ATP.

To synthesize fatty acids with more than 16 carbon atoms, the palmitic acid from this biosynthesis must enter a fatty acid elongation pathway in either the mitochondria or the microsomes.

FATTY ACID ELONGATION

Unlike cytoplasmic fatty acid synthesis, the mitochondria do not use malonate to elongate fatty acids. Instead, acetyl-CoA units are added successively to lengthen fatty acids to 18 to 24 carbon atoms.

In the microsomes, saturated and unsaturated fatty acids are lengthened by adding malonyl-CoA units. Here, the monounsaturated fatty acids, oleic acid and palmitoleic acid, are synthesized by cytochrome-b_5 oxygenase in the cytoplasm.

ESSENTIAL FATTY ACIDS

Since saturated, monounsaturated, and selected polyunsaturated fatty acids may be synthesized in humans, man can compensate for their absence from the diet. Several polyunsaturated fatty acids, however, are essential in the diet because they cannot be synthesized. These include *linoleic* and *linolenic acids*. Arachidonic acid may be synthesized from linoleic and linolenic acids; hence, though it is required, it is not considered "essential." The most common case of essential fatty acid deficiency is seen in the patient on long-term intravenous hyperalimentation who is not receiving intravenous fat solutions.

Problems 10–16

Match cytoplasmic fatty-acid biosynthesis and β-oxidation to the features described in Problems 10–16:

A. Cytoplasmic synthesis of saturated fatty acids
B. β-Oxidation of saturated fatty acids

10. Requires ACP to transfer the fatty-acid chain.
11. Uses $NAD^+ \rightarrow NADH$.
12. Malonyl-CoA is an essential precursor.
13. Can involve 18-carbon to 24-carbon fatty acids.
14. Uses $FAD \rightarrow FADH_2$.

15. Employs NADPH $\rightarrow$ NADP$^+$.

16. Can be stimulated by citrate.

Problems 17–21

Draw the missing structures in these reactions of cytoplasmic fatty-acid synthesis. Describe the reactions and name the enzyme type (i.e., synthase, transferase, hydratase, or whatever):

17.
$$^-OOC-CH_2-\overset{\displaystyle O}{\overset{\|}{C}}-S-CoA + ACP\text{-}SH \rightleftharpoons \cdots$$

18.
$$CH_3-CH=CH-\overset{\displaystyle O}{\overset{\|}{C}}-S-ACP + NADPH + H^+ \rightleftharpoons \cdots$$

19.
$$\cdots \longrightarrow CH_3-\overset{\displaystyle O}{\overset{\|}{C}}-CH_2-\overset{\displaystyle O}{\overset{\|}{C}}-S-ACP + CO_2 + synthase\text{-}SH$$

20.
$$CH_3-\overset{\displaystyle OH}{\overset{|}{C}H}-CH_2-\overset{\displaystyle O}{\overset{\|}{C}}-S-ACP \rightleftharpoons H_2O + \cdots$$

21.
$$CH_3-CH_2-CH_2-\overset{\displaystyle O}{\overset{\|}{C}}-S\text{-synthase} + {}^-OOC-CH_2-\overset{\displaystyle O}{\overset{\|}{C}}-S-ACP \longrightarrow \cdots$$

KETONE-BODY METABOLISM

Acetoacetic acid, β-hydroxybutyric acid, and acetone are classified as *ketone bodies*. The term "ketone bodies" is inaccurate, since β-hydroxybutyrate lacks a keto group.

Acetoacetic acid is the principal ketone body synthesized by the liver mitochondria. As shown in Figure 12-4, acetate from acetyl-CoA is dimerized to yield acetoacetyl-CoA. This CoA, however, cannot be readily removed to produce acetoacetic acid. Instead, another acetyl-CoA must first add to the acetoacetyl-CoA to yield the six-carbon intermediate, β-hydroxy-β-methyl-glutaryl-CoA (HMG-CoA). Acetyl-CoA is then removed from HMG-CoA in the liver to liberate acetoacetic acid. β-Hydroxybutyrate dehydrogenase reduces much of this acetoacetic acid to β-hydroxybutyric acid. In addition, a decarboxylase converts some of this acetoacetate to acetone. Acetone is metabolized very slowly. Because of its volatility, most of the acetone evaporates through the lung alveoli.

Extrahepatic tissues, such as skeletal muscle and heart muscle, utilize the two ketone bodies other than acetone as a fuel (acetone cannot be significantly degraded). These tissues oxidize β-hydroxybutyrate to acetoacetate and then add CoA-SH by either of two different routes to create acetoacetyl-CoA, as shown in Figure 12-5. Finally, they cleave acetoacetyl-CoA into two acetyl-CoA molecules, which can enter the TCA cycle.

High serum levels of acetoacetate and β-hydroxybutyrate constitute *ketonemia*. Normal urine lacks ketone bodies. The common hospital tests for *ketonuria*, or ketone bodies in the urine, measure only acetone and acetoacetate. Such tests may fail to detect ketonuria if β-hydroxybutyrate predominates.

The combination of ketonemia and ketonuria is termed *ketosis*. Ketosis occurs whenever the rate of hepatic ketone-body production exceeds the rate of

Figure 12-4 *Ketone body synthesis in the liver.*

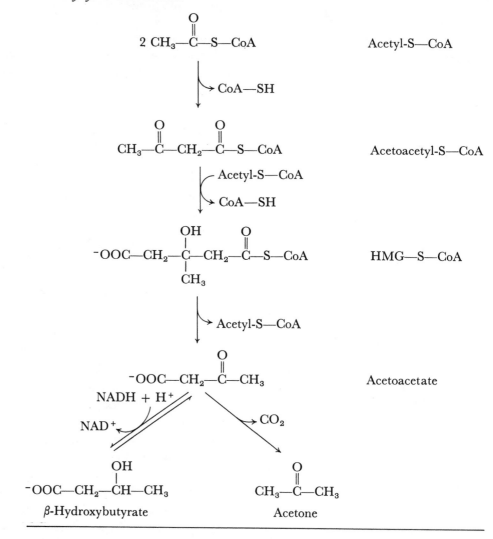

peripheral utilization. The liver overproduces ketone bodies during severe carbohydrate deficiency for at least two reasons. First, carbohydrate deficiency depletes the TCA-cycle intermediates and slows the entrance of acetyl-CoA into this cycle. Second, acetyl-CoA carboxylase, the rate-controlling enzyme of fatty-acid synthesis, is inhibited by the absence of citrate, thereby blocking another route of acetyl-CoA metabolism. Thus, acetyl-CoA accumulates in the liver and is excessively converted to ketone bodies. Severe ketonemia overwhelms the blood buffers, causing metabolic acidosis with an increased anion gap (the "unmeasured anions" in this case are β-hydroxybutyrate and acetoacetate). Severe carbohydrate deficiency is the mechanism underlying diabetic ketoacidosis, alcoholic ketoacidosis, and starvation ketosis.

The Food and Nutrition Board of the United States recommends that the adult diet contain at least 100 g or 400 Cal of carbohydrate daily to generate enough oxaloacetate to maintain the TCA cycle and prevent ketosis. In addition, carbohydrate deficiency causes protein wasting, because much of the dietary protein is converted via deamination and gluconeogenesis to glucose.

Figure 12-5

Ketone-body degradation in extrahepatic tissues.

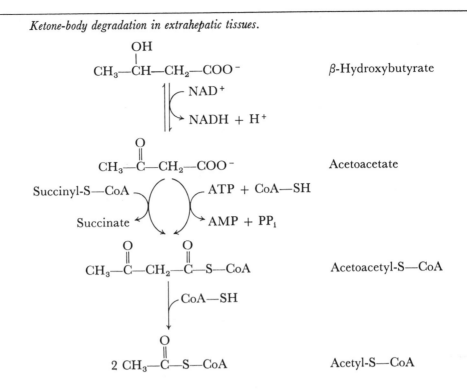

$$
\underset{\text{β-Hydroxybutyrate}}{\overset{\displaystyle \overset{\text{OH}}{|}}{CH_3—CH—CH_2—COO^-}}
$$

NAD⁺ ↷

NADH + H⁺

$$
\underset{\text{Acetoacetate}}{CH_3—\overset{\displaystyle \overset{O}{\|}}{C}—CH_2—COO^-}
$$

Succinyl-S—CoA   ATP + CoA—SH

Succinate   AMP + PP$_i$

$$
\underset{\text{Acetoacetyl-S—CoA}}{CH_3—\overset{\displaystyle \overset{O}{\|}}{C}—CH_2—\overset{\displaystyle \overset{O}{\|}}{C}—S—CoA}
$$

CoA—SH

$$
\underset{\text{Acetyl-S—CoA}}{2\ CH_3—\overset{\displaystyle \overset{O}{\|}}{C}—S—CoA}
$$

CHOLESTEROL BIOSYNTHESIS

Man has two sources of cholesterol: dietary cholesterol and its de novo synthesis from acetate. The greater the dietary intake of cholesterol, the lower the rate of cholesterol biosynthesis in the liver and adrenal cortex. Although cholesterol itself has no caloric value in foods, its presence in the diet spares the energy needed to synthesize cholesterol.

The first two steps of cholesterol biosynthesis are the same as those in ketone-body synthesis; acetyl-CoA is dimerized to acetoacetyl-CoA, to which another acetyl-CoA is added to create β-hydroxy-β-methylglutaryl-S-CoA (HMG-S-CoA).

HMG-S-CoA reductase then catalyzes the committed and rate-controlling step of cholesterol biosynthesis, the reduction of the —C(=O)—S—CoA in HMG-S-CoA to —CH$_2$OH in mevalonic acid, as shown in Figure 12-6. Fasting reduces cholesterol synthesis by reducing the synthesis of this reductase enzyme.

Next, mevalonic acid is phosphorylated three times with ATP to create mevalonate-3-phospho-5-pyrophosphate, which loses its carboxyl and 5-pyrophosphate groups to become the five-carbon intermediate, isopentenyl pyrophosphate. An isomerase changes the position of the carbon-carbon double bond in this intermediate to yield another five-carbon intermediate, 3,3-dimethylallyl pyrophosphate.

These two five-carbon intermediates then condense to create the 10-carbon compound, geranyl pyrophosphate. To this, another five-carbon isopentenyl pyrophosphate is added to produce the 15-carbon compound, farnesyl pyrophosphate, as shown in Figure 12-7.

Two farnesyl pyrophosphate molecules dimerize to form presqualene pyrophosphate, which is then reduced to the 30-carbon compound, squalene.

Figure 12-6

Cholesterol biosynthetic pathway from acetyl-CoA to geranyl pyrophosphate.

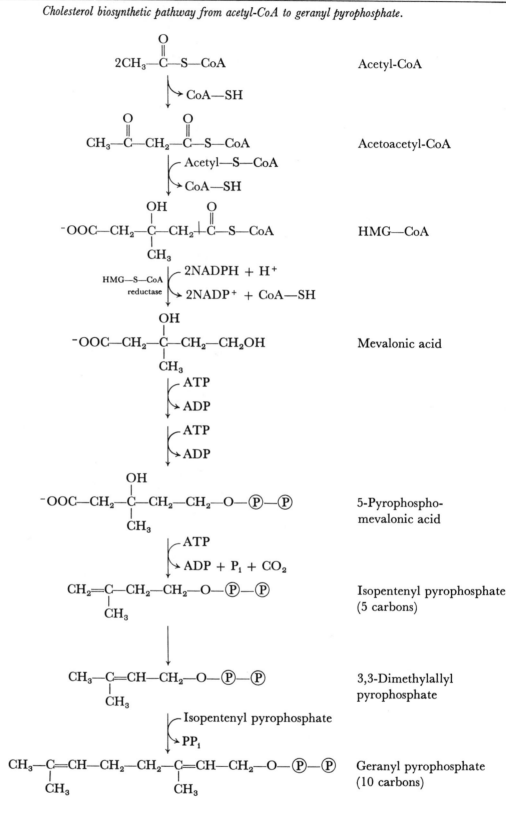

Figure 12-7   *Cholesterol biosynthetic pathway from geranyl pyrophosphate to cholesterol.*

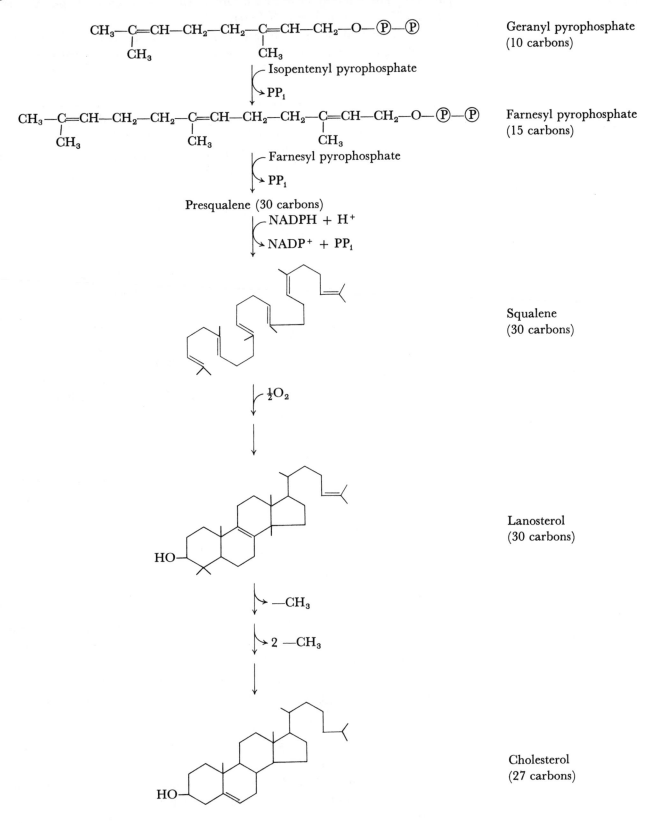

CH₃—C=CH—CH₂—CH₂—C=CH—CH₂—O—Ⓟ—Ⓟ   Geranyl pyrophosphate
       | | (10 carbons)
      CH₃   CH₃

        Isopentenyl pyrophosphate
        PP₁

CH₃—C=CH—CH₂—CH₂—C=CH—CH₂—CH₂—C=CH—CH₂—O—Ⓟ—Ⓟ   Farnesyl pyrophosphate
      | | | (15 carbons)
     CH₃   CH₃   CH₃

        Farnesyl pyrophosphate
        PP₁

Presqualene (30 carbons)
        NADPH + H⁺
        NADP⁺ + PP₁

Squalene
(30 carbons)

½O₂

Lanosterol
(30 carbons)

—CH₃

2 —CH₃

Cholesterol
(27 carbons)

151

Oxygenation is followed by cyclization to yield lanosterol, the first sterol of the pathway. Three methyl groups are subsequently removed to shape this sterol into the 27-carbon product, cholesterol.

The principal cause of cholesterol gallstones is an increase in the biliary ratio of cholesterol to bile salts. Bile salts and phosphoglycerides normally solubilize biliary cholesterol. In bile-salt deficiency, this cholesterol precipitates.

PHOSPHO-GLYCERIDE SYNTHESIS

Absent or defective membrane receptors for LDL (low-density lipoprotein) leads to massive overproduction of cholesterol in familial hypercholesterolemia.

Cytidine triphosphate (CTP) is employed to attach phosphoethanolamine to diglycerides to create phosphatidylethanolamine:

$$CTP + phosphoethanolamine \rightleftharpoons CDP\text{-}ethanolamine + PP_i$$

$$CDP\text{-}ethanolamine + diglyceride \rightleftharpoons phosphatidylethanolamine + CMP$$

The synthesis of phosphatidylcholine in the lungs of newborns is a critical factor in enabling the postnatal expansion of the lung, since phosphatidylcholine serves as a pulmonary surfactant.

In humans, there are two pathways for synthesizing phosphatidylcholine. In the first pathway, CTP combines with phosphocholine to create CDP-choline, which donates its choline group to a diglyceride to generate phosphatidylcholine. In the second, each of three S-adenosylmethionine molecules donates its methyl group to the ethanolamine of phosphatidylethanolamine to convert it to trimethylethanolamine, or choline:

$$Phosphatidylethanolamine + 3 \; S\text{-}adenosylmethionine \longrightarrow$$
$$phosphatidylcholine + 3 \; S\text{-}adenosylhomocysteine$$

$$\overset{OH}{\underset{|}{CH_2}}\!\!-\!CH_2\!-\!{}^+NH_3$$

Ethanolamine

$$\overset{OH}{\underset{|}{CH_2}}\!\!-\!CH_2\!-\!{}^+\overset{CH_3}{\underset{CH_3}{N}}\!-\!CH_3$$

Choline

SPHINGOLIPID CATABOLISM

As indicated in Figure 12-8, a variety of hydrolytic enzymes are required to catabolize sphingolipids. The absence of any one of these enzymes leads to a sphingolipid-deposition disease known as a *sphingolipidosis*.

In Niemann-Pick disease, the lack of sphingomyelinase prevents the removal of phosphocholine from ceramide in sphingomyelin. Thus, sphingomyelin accumulates within the brain, liver, and spleen.

Absence of hexosaminidase A in Tay-Sachs disease blocks the cleavage of N-acetylgalactosamine (a hexosamine) from GM_2 gangliosides, which deposit in the brain and in the macula of the retina, causing cherry-red spots.

In metachromatic leukodystrophy, the absence of a sulfatidase, arylsulfatase A, permits sulfatides to accumulate in the white matter of the brain.

β-Glucosidase deficiency in Gaucher's disease prevents the cleavage of

Figure 12-8 — Metabolic diseases characterized by inabilities to degrade sphingolipids. Cer = ceramide; NAcNA = N-acetylneuraminic acid; Gal = galactose; Glc = glucose; PChol = phosphorylcholine; NAcGal = N-acetylgalactosamine. (Reproduced by permission. From R. W. Albers, G. J. Siegel, R. Katzman, and B. W. Agranoff, Basic Neurochemistry. Boston: Little, Brown, 1972. Fig. 23-1.)

Disease	Major sphingolipid accumulated	Enzyme defect
Niemann-Pick	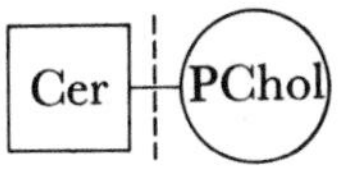 Cer—PChol — Sphingomyelin	Sphingomyelinase
Gaucher	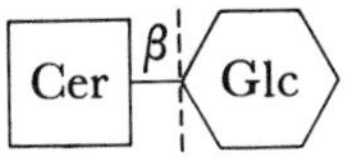 Cer—β—Glc — Ceramide glucoside (glucocerebroside)	β-Glucosidase
Krabbe	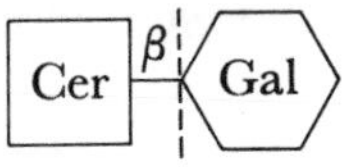 Cer—β—Gal — Ceramide galactoside (galactocerebroside)	β-Galactosidase
Metachromatic Leukodystophy	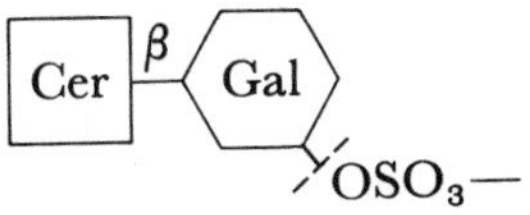 Cer—β—Gal—OSO₃— — Ceramide galactose-3-sulfate (sulfatide)	Sulfatidase
Ceramide Lactoside Lipidosis	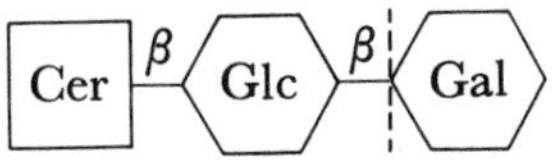 Cer—β—Glc—β—Gal — Ceramide lactoside	β-Galactosidase
Fabry	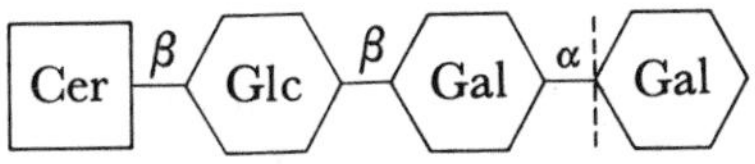 Cer—β—Glc—β—Gal—α—Gal — Ceramide trihexoside	α-Galactosidase
Tay-Sachs	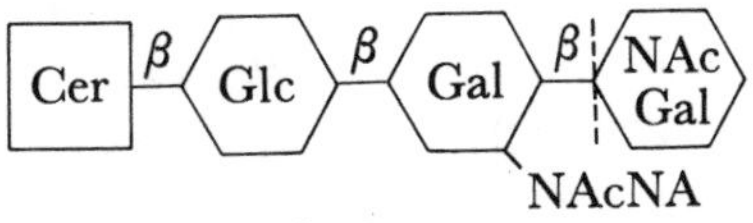 Cer—β—Glc—β—Gal—β—NAcGal (NAcNA) — Ganglioside GM₂	Hexosaminidase A
Tay-Sachs Variant	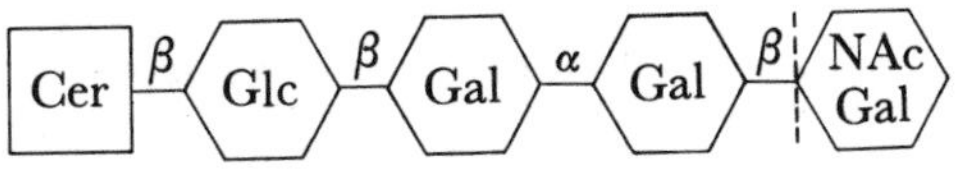 Cer—β—Glc—β—Gal—α—Gal—β—NAcGal — Globoside (plus Ganglioside GM₂)	Total Hexosaminidase
Generalized Gangliosidosis	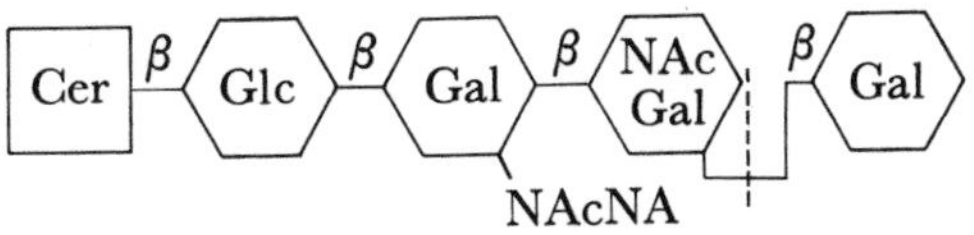 Cer—β—Glc—β—Gal—β—NAcGal (NAcNA)—β—Gal — Ganglioside GM₁	β-Galactosidase

glucose from glucocerebrosides, which deposit in the liver, the spleen, and, in infants, the brain. The bone marrow exhibits cerebroside-filled Gaucher cells.

Absence of α-galactosidase in Fabry's disease causes ceramide-trisaccharides to accumulate in the skin and kidneys.

HORMONAL CONTROL OF LIPID METABOLISM

Insulin promotes triglyceride synthesis by two mechanisms. First and foremost, it hastens glucose entry into adipose tissue, thereby providing fuel for fatty-acid and triglyceride synthesis. Second, insulin inhibits the hormone-sensitive lipase in adipose tissue by promoting its dephosphorylation, which renders the lipase inactive. Thus, insulin is an *antilipolytic* hormone.

Lipolytic hormones stimulate adenylcyclase in adipose tissues to convert ATP to cyclic AMP. Cyclic AMP activates protein kinase, which in turn phosphorylates hormone-sensitive lipase. This activated lipase hydrolyzes triglycerides and releases a surge of albumin-bound, fatty acids into the blood. This cyclic-AMP-mediated control of lipolysis in adipose tissue resembles cyclic-AMP-induced glycogenolysis in the liver.

The most potent lipolytic hormone is epinephrine. The plasma free-fatty-acid level rises markedly after an injection of epinephrine. Other lipolytic hormones include growth hormone (STH), thyroid hormones, glucocorticoids, and glucagon.

Hyperthyroidism makes the hormone-dependent lipase more sensitive to epinephrine. Hypercholesterolemia (high serum cholesterol levels) occurs in *hypothyroidism*, because of impaired removal of cholesterol from the blood.

Glucocorticoid deficiency in adrenocortical hypofunction reduces the sensitivity of hormone-sensitive lipase to lipolytic agents. *Glucocorticoid excess, or Cushing's syndrome*, causes an unexplained redistribution of fat from the extremities to the trunk, face ("moon face"), and lower neck posteriorly ("buffalo hump").

Immediately after a meal, insulin is released and the lipolytic-hormone levels diminish, allowing insulin to promote fatty-acid, triglyceride, and glycogen synthesis. Insulin, therefore, acts as a fuel-storing hormone. Several hours later, insulin begins to disappear from the bloodstream, and the lipolytic, glycogenolytic, and gluconeogenic hormones start to mobilize triglycerides, glycogen, and proteins, respectively. During fasting, these fuel-mobilizing hormones predominate and insulin almost vanishes. The lack of the insulin effect in diabetes mellitus leads to excess lipid mobilization, which results often in hypertriglyceridemia.

Problem 22

An Antarctic explorer decides to use butter (9 Cal/gram) rather than carbo-hydrate (4 Cal/gram) as his calorie source during an expedition. His protein source, dried beef, has virtually no carbohydrate. He takes multiple vitamin pills. After two weeks of this diet, he will have:

A. protein sparing
B. normal metabolism
C. ketosis
D. a normally functioning TCA cycle

| Problems 23–26 | Match the incomplete reactions of cholesterol biosynthesis below to their descriptions in Problems 23–26: |

A. Lanosterol → cholesterol
B. HMG-CoA → mevalonic acid + CoA-SH
C. Acetoacetyl-CoA + acetyl-CoA → HMG-CoA + CoA-SH
D. Squalene → lanosterol

23. Sterol rings created from a branched-chain structure.
24. Loss of three methyl groups.
25. Rate-controlling step of cholesterol biosynthesis.
26. Same reaction used in ketone-body synthesis.

| Problem 27 | Cholesterol is *not* the precursor of human: |

A. bile acids   C. vitamin D
B. steroid hormones   D. α-tocopherol

| Problem 28 | CH_3—C(=O)—S— synthase condenses with radioactively labeled ^-OOC—*CH_2—C(=O)—S—ACP in fatty-acid biosynthesis to yield butyryl-S-ACP after several reactions. Where will the radioactive label appear? |

A.
$$*CH_3-CH_2-CH_2-\overset{\overset{\displaystyle O}{\|}}{C}-S-ACP + CO_2$$

B.
$$CH_3-*CH_2-CH_2-\overset{\overset{\displaystyle O}{\|}}{C}-S-ACP + CO_2$$

C.
$$CH_3-CH_2-*CH_2-\overset{\overset{\displaystyle O}{\|}}{C}-S-ACP + CO_2$$

D.
$$CH_3-CH_2-CH_2-*\overset{\overset{\displaystyle O}{\|}}{C}-S-ACP + CO_2$$

E.
$$CH_3-CH_2-CH_2-\overset{\overset{\displaystyle O}{\|}}{C}-S-ACP + *CO_2$$

| Problem 29 | If the malonyl-ACP were labeled as $^-OO*C$—CH_2—C(=O)—S—ACP, where would this label appear in the above products? |

| Problem 30 | How many malonate units are consumed in the cytoplasmic synthesis of palmitic acid (16 carbons)? |

A. 5   C. 7
B. 6   D. 8

Problem 31

Choose the *incorrect* statement about the lipolytic hormones:

A. They lead to phosphorylation of hormone-sensitive lipase.
B. They activate adenylcyclase in adipose tissue.
C. They lower serum cholesterol levels.
D. They act using cyclic AMP as a second messenger.

Problem 32

During diabetic ketoacidosis, the extrahepatic tissues metabolize acetoacetic acid initially to:

A. HMG-CoA
B. acetoacetyl-CoA

C. cholesterol
D. β-hydroxybutyric acid

Problems 33–35

Match the sphingolipid hydrolytic enzymes below to the components that they cleave:

A. Sphingomyelinase
B. Hexosaminidase
C. β-Galactosidase

33. *N*-Acetylgalactosamine
34. Phosphocholine
35. Galactose

Problem 36

Choose the two essential fatty acids from the list below. How do they differ from the other fatty acids listed?

A. Stearic acid
B. Oleic acid
C. Linoleic acid

D. Palmitoleic acid
E. Linolenic acid

Problem 37

The methyl donor in the synthesis of phosphatidylcholine from phosphatidylethanolamine is:

A. Methyl-THFA
B. *S*-Adenosylmethionine
C. Methyl-B_{12}
D. DA-cobalamin

ANSWERS

1. D. (Biotin is required in most carboxylations, except for those of amino acids; see Ch. 5.)
2. B. (Methylmalonyl-CoA mutase requires DA-B_{12}; see Ch. 5. Methylmalonic acid accumulates in vitamin B_{12} deficiency.)
3. B, C. (Glutamate, having five carbons, could not arise from the transamination of succinate, which has four carbons.)

4.

$$R—\overset{\overset{\displaystyle O}{\|}}{C}—S—CoA \ + \ CH_3—\overset{\overset{\displaystyle O}{\|}}{C}—S—CoA$$

Cleavage reaction using the SH binding site of CoA; hence, this is a thiolytic cleavage. A thiolase catalyzes this final reaction of β-oxidation.

5.

$$R—CH{=}CH—\overset{\overset{\displaystyle O}{\|}}{C}—S—CoA \ + \ H_2O$$

Hydration of a carbon-carbon double bond catalyzed by a hydratase.

6.

$$R—CH{=}CH—\overset{\overset{\displaystyle O}{\|}}{C}—S—CoA \ + \ FADH_2$$

Dehydrogenation reaction catalyzed by acyl-CoA dehydrogenase; this is the initial reaction of β-oxidation.

7.

$$R—\overset{\overset{\displaystyle O}{\|}}{C}—CH_2—\overset{\overset{\displaystyle O}{\|}}{C}—S—CoA \ + \ NADH \ + \ H^+$$

Dehydrogenation reaction catalyzed by β-hydroxyacyl-CoA dehydrogenase.

8. A. One mole ATP.
    B. Six passes, not seven. (The sixth passage converts butyric acid, a four-carbon fatty acid, into two moles acetyl-CoA.)
    C. Six moles $FADH_2$ and six moles NADH.
    D. ATP produced = $6FADH_2 \times 2ATP/FADH_2 + 6NADH \times 3ATP/NADH$ = 30 moles ATP
        One mole of ATP was consumed in the initial activation step; therefore, the net ATP yield = $30 - 1 = 29$ moles.
        Six moles O_2 are consumed in the reoxidation of $FADH_2$ and NADH ($\frac{1}{2}O_2/FADH_2$ and $\frac{1}{2}O_2/NADH$).

9. A. Seven moles GTP.
    B. Seven moles $FADH_2$ and 21 moles NADH.
    C. ATP yield = 12 moles ATP per turn of TCA cycle $\times$ seven turns = 84 moles ATP
        Fourteen moles O_2 are consumed in the reoxidation of $FADH_2$ and NADH from the TCA cycle.
    D. Net ATP yield = $29 + 84 = 113$ moles

$$RQ = \frac{14 \text{ moles } CO_2 \text{ produced}}{20 \text{ moles } O_2 \text{ consumed}} = 0.70$$

10. A.
11. B.
12. A.
13. B. (Cytoplasmic fatty-acid biosynthesis alone cannot generate fatty acids with more than 16 carbons.)
14. B.
15. A.
16. A.

17.

$$^-OOC—CH_2—\overset{\overset{\displaystyle O}{\|}}{C}—S—ACP \ + \ CoA\text{-}SH$$

Malonate is transferred from CoA to ACP. The enzyme is a transferase.

18.

$$CH_3\!-\!CH_2\!-\!CH_2\!-\!\overset{\displaystyle O}{\overset{\displaystyle \|}{C}}\!-\!S\!-\!ACP + NADP^+$$

A reductase hydrogenates the carbon-carbon double bond of crotonyl-S-ACP to form butyryl-S-ACP.

19.

$$CH_3\!-\!\overset{\displaystyle O}{\overset{\displaystyle \|}{C}}\!-\!S\!-\!synthase + {}^-OOC\!-\!CH_2\!-\!\overset{\displaystyle O}{\overset{\displaystyle \|}{C}}\!-\!S\!-\!ACP$$

The condensation of acetyl-S-synthase with malonyl-S-ACP is catalyzed by β-ketoacyl-ACP synthase.

20.

$$CH_3\!-\!CH\!=\!CH\!-\!\overset{\displaystyle O}{\overset{\displaystyle \|}{C}}\!-\!S\!-\!ACP$$

β-Hydroxylbutyryl-S-ACP is dehydrated to form crotonyl-S-ACP. The enzyme, enoyl hydratase, is named for the reverse reaction.

21.

$$CH_3\!-\!CH_2\!-\!CH_2\!-\!\overset{\displaystyle O}{\overset{\displaystyle \|}{C}}\!-\!CH_2\!-\!\overset{\displaystyle O}{\overset{\displaystyle \|}{C}}\!-\!S\!-\!ACP + CO_2 + synthase\text{-}SH$$

Butyryl-S-synthase condenses with malonyl-S-ACP to form a six-carbon, β-keto, fatty acyl-S-ACP. The reaction is catalyzed by β-ketoacyl-ACP synthase.

22. C. (Such a carbohydrate deficiency leads to protein wasting, ketosis, and depletion of TCA-cycle intermediates.)

23. D.

24. A.

25. B.

26. C.

27. D. (Some cholesterol becomes 7-dehydrocholesterol, which is converted by irradiation, i.e., through sunlight on the skin, to cholecalciferol, or vitamin D_3.)

28. C.

29. E.

30. C. (Seven malonate units, with 21 carbons total, give 14 carbons to palmitic acid and liberate seven moles CO_2. The remaining two carbons of palmitic acid come from the initial acetate.)

31. C.

32. B.

33. B.

34. A.

35. C.

36. C, E. (Stearic acid is saturated. Oleic and palmitoleic acids are monounsaturated; hence, they are synthesized in humans. Humans, however, cannot synthesize linoleic acid with two double bonds nor linolenic acid with three.)

37. B.

REFERENCES

Brown, M. S., and Goldstein, J. L.   Familial hypercholesterolemia: a genetic defect in the low-density lipoprotein receptor. *N. Eng. J. Med.* 294:1386, 1976.

Gelehrter, T.   Enzyme induction. *N. Engl. J. Med.* 294:522, 1976.

Goodman, L. S., and Gilman, A.   *The Pharmacological Basis of Therapeutics*

(5th ed.). New York: Macmillan, 1975. Pp. 774–751 (lipid-lowering drugs), 1378–1379 (STH), 1407 (thyroid hormones), 1482 (glucocorticoids), 1515–1516 (insulin).

Krumdieck, C. L.   Intestinal regulation of hepatic cholesterol synthesis: an hypothesis. *Am. J. Clin. Nutr.* 30:255, 1977.

Lehninger, A. L.   *Biochemistry: The Molecular Basis of Cell Structure and Function* (2nd ed.). New York: Worth, 1975. Pp. 543–557, 659–689.

Quarfordt, S. H.   Methods for the in vivo estimation of human cholesterol dynamics. *Am. J. Clin. Nutr.* 30:967, 1977.

Siegel, G. J., Albers, R. W., Katzman, R., and Agranoff, B. W.   *Basic Neurochemistry* (2nd ed.). Boston: Little, Brown, 1976. Pp. 308–328.

White, A., Handler, P., and Smith, E. L.   *Principles of Biochemistry* (5th ed.). New York: McGraw-Hill, 1973. Pp. 542–601.

13 Amino Acid Metabolism and Biosynthesis

PROTEOLYSIS

Proteolysis, or the hydrolysis of the peptide bonds of dietary protein, begins in the stomach, where gastric HCl acidifies the food to pH 2 to 3, the optimum pH for the proteolytic enzyme, pepsin. Since HCl in the stomach is too dilute to hydrolyze peptide bonds readily by itself, gastric proteolysis is carried out primarily by pepsin. On entering the intestine, the oligopeptides and polypeptides produced by pepsin digestion encounter the pancreatic proteases—i.e., trypsin, chymotrypsin, and carboxypeptidase—and are cleaved to a mixture of free amino acids and oligopeptides.

The gastric and pancreatic proteases are secreted as zymogens, or inactive precursors, termed pepsinogen, trypsinogen, and chymotrypsinogen. Gastric HCl activates pepsin, and enterokinase, an intestinal enzyme, activates trypsinogen to trypsin. Trypsin, in turn, activates chymotrypsinogen.

The intestinal mucosa contains and also secretes the protease leucine aminopeptidase and the dipeptidases, which further degrade oligopeptides into free amino acids. The small intestine then absorbs these free amino acids by active transport. They enter the hepatic portal-venous system and are carried to the main organ of amino acid metabolism, the liver.

AMINO ACID CATABOLISM

Amino acids in the blood readily penetrate the renal glomeruli, but they do not normally appear in the urine in appreciable amounts, because the renal tubules avidly reabsorb them. In cystinuria, an inherited disorder, the kidney tubules fail to reabsorb the basic amino acids cystine, ornithine, arginine, and lysine (the mnemonic is COAL). In this and other aminoacidurias, the urine drains significant quantities of amino acids from the blood.

The main site of amino acid catabolism is the liver. The kidney is a secondary site of catabolism (i.e., via metabolic conversion, rather than through loss into the glomerular filtrate). Amino-acid catabolism generally begins with the removal of the α-amino group, leaving an α-keto acid. The liver converts most of the ammonium (NH_4^+) ions, produced by deamination, to urea, which is secreted into the urine. Some of these ammonium ions are secreted directly into the urine by the kidneys. The carbon skeleton of the α-keto acid is then degraded to useful intermediates or metabolic endproducts, such as oxalate.

Protein catabolism generates 4 Cal/gram, the same energy yield as in carbohydrate breakdown. Dietary protein has a high *specific dynamic action*, or post-absorptive thermogenesis, which is defined as the quantity of heat, relative to the basal state, released during catabolism following the ingestion of food.

Removal of Alpha-Amino Groups

The principal means for removing α-amino groups is a transamination reaction with α-ketoglutarate that yields the corresponding α-keto acid plus glutamate, as shown in Figure 13-1. (Since glutarate sounds like glutamate, it is easy to remember that they are an α-keto acid and α-amino acid pair.)

Figure 13-1

Transamination followed by oxidative deamination.

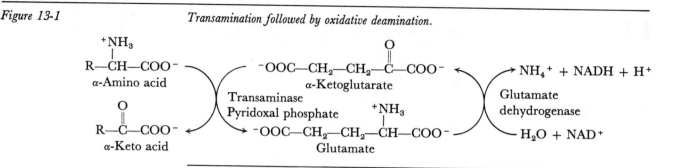

Pyridoxal phosphate, a member of the vitamin-B_6 group, is an essential coenzyme in transaminations, as well as in many other amino acid reactions. It may accept an amino group to become pyridoxamine phosphate.

After the transaminases have gathered most of the α-amino groups into glutamate, the enzyme glutamate dehydrogenase liberates the amino groups as ammonia and regenerates α-ketoglutarate. This is a deamination, rather than a transamination, reaction. Furthermore, it is an oxidative deamination, because the NAD^+ in this reaction is reduced to NADH, which is reoxidized by the electron-transport chain. Glutamate dehydrogenase can, alternatively, utilize $NADP^+$.

The second method for removing α-amino groups employs oxidative de-amination without transamination. This pathway, of minor importance in human metabolism, uses FMN-linked amino-acid oxidase. The $FMNH_2$ produced then reduces O_2 to H_2O_2. Human catalase degrades this hydrogen peroxide to $H_2O + \frac{1}{2}O_2$:

$$\alpha\text{-Amino acid} + FMN + H_2O \longrightarrow \alpha\text{-keto acid} + NH_3 + FMNH_2$$
$$FMNH_2 + O_2 \longrightarrow FMN + H_2O_2$$
$$H_2O_2 \longrightarrow H_2O + \tfrac{1}{2}O_2$$

The third mechanism for removing α-amino groups is by nonoxidative deamination. This is the main route for removing the α-amino groups from asparagine, cysteine, glutamine, histidine, and serine.

Renal Ammonia Excretion

Simpson (1971) states that the kidneys excrete 60% of the daily load of hydrogen ions from metabolism as ammonium ions and 40% as titratable acids, chiefly phosphate.

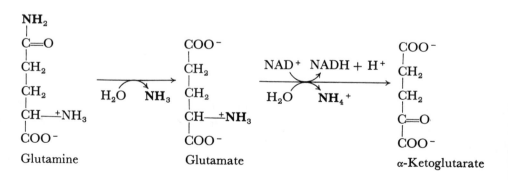

Glutaminase nonoxidatively deaminates glutamine to glutamate in the kidneys. Glutamate dehydrogenase then oxidatively deaminates glutamate to α-ketoglutarate. Two NH_3 molecules are secreted into the urine for every glutamine twice-deaminated to α-ketoglutarate. Once inside the renal tubules, NH_3 is protonated to NH_4^+, which cannot leave the tubules, because of ion-trapping.

A second pathway transaminates (rather than deaminates) glutamine and then deaminates the product to yield α-ketoglutarate plus ammonia.

Many studies have disputed the often-repeated statement that ammonia is highly toxic, particularly to the brain. Merino et al. (1975), for instance, infused NH_4Cl intravenously into normal dogs. They found that coma developed only after reaching the stupendously high serum ammonia levels of 2.0 to 3.5 g per 100 ml (1000-fold above normal). The moderately elevated serum ammonia level in hepatic coma and in Reye's syndrome does not itself cause neurologic abnormalities. Fischer et al. (1974) have shown that a distortion in the serum levels of amino acids occurs in hepatic coma, but whether these amino acids play a part in causing hepatic coma is still uncertain.

Urea Cycle

Renal ammonia excretion accounts for only a small fraction of the daily loss of the nitrogen resulting from amino-acid catabolism. The vast majority of nitrogen excreted is in the form of urea, which penetrates the renal glomeruli and is lost in the urine. The urea cycle occurs predominately in the liver, although it also exists in the brain and kidneys. Many tissues possess all but one urea-cycle enzyme, and they use this pathway to synthesize arginine. Their lack of arginase, however, prevents urea formation from the arginine.

Carbamoyl-phosphate synthetase generates the unstable intermediate, carbamoyl phosphate, from CO_2, NH_3, and H_2O, expending two moles of ATP in the reaction. The glutamate-dehydrogenase reaction supplies the required NH_3. Another carbamoyl-phosphate synthetase exists in the cytoplasm to generate carbamoyl phosphate for pyrimidine synthesis.

On entering the urea cycle, this carbamoyl group, $—C(=O)—NH_2$, is added by ornithine carbamoyltransferase, inside the mitochondria, to the basic amino acid, ornithine, to create citrulline, as shown in Figure 13-2. Citrulline diffuses into the cytoplasm, which is the reaction site for the remaining three urea-cycle enzymes.

Argininosuccinate synthetase in the cytoplasm condenses citrulline and aspartate to generate argininosuccinate. In the process, ATP is split to $AMP + PP_i$ (the pyrophosphate is then hydrolyzed to two moles P_i).

Argininosuccinate lyase cleaves argininosuccinate to arginine and fumarate; it transforms a single bond in succinate to the double bond of fumarate. This fumarate enters the TCA cycle.

Many tissues possess the enzymes needed to synthesize arginine by this pathway. Only the liver, kidneys, and brain have the arginase to hydrolyze arginine to urea and ornithine. Ornithine completes the urea cycle by returning to the mitochondria, where it again condenses with carbamoyl phosphate.

One nitrogen atom of the urea produced comes from glutamate via carbamoyl phosphate, while the other is taken from aspartate in the argininosuccinate-synthetase step. The carbon atom of the urea is derived from CO_2, also via carbamoyl phosphate.

Figure 13-2   *The urea cycle.*

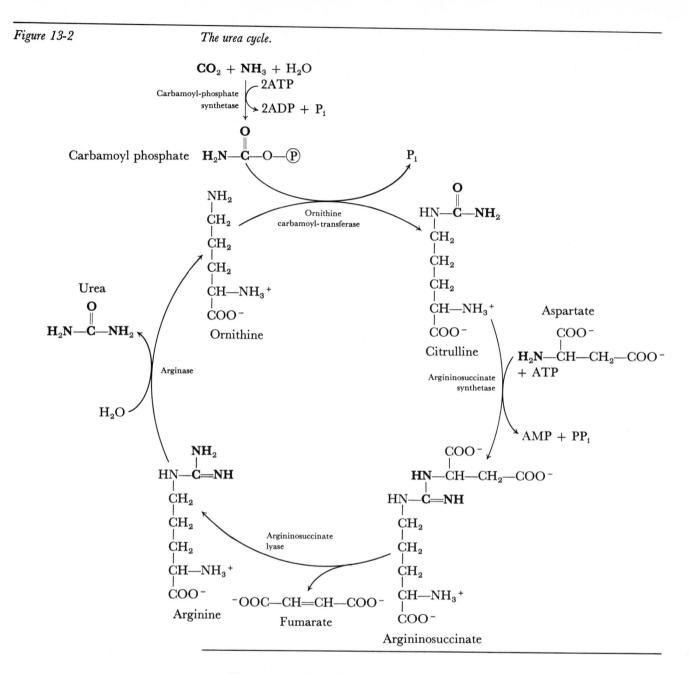

The net reaction of the urea cycle consumes the equivalent of four moles ATP; two moles ATP must be expended to convert the AMP produced in the argininosuccinate-synthetase step to ATP. The reaction may be written as:

$$CO_2 + 2NH_3 + 3ATP + 3H_2O \longrightarrow urea + 2ADP + AMP + 4P_i$$

For each enzyme of the urea cycle, there is a rare, inherited, deficiency disorder that is almost always fatal in infancy. The blood urea-nitrogen concentration (BUN) is extremely low in these cases, which is due to impaired urea synthesis.

Problems 1–5 Draw the missing structures in the reactions below and name all the structures and enzymes:

1.
$$H_2N-\overset{\overset{\displaystyle O}{\|}}{C}-CH_2-CH_2-\overset{\overset{\displaystyle +NH_3}{|}}{CH}-COO^- + H_2O \longrightarrow NH_3 + \cdots$$

2. $CO_2 + NH_3 + H_2O + 2ATP \longrightarrow 2ADP + P_i + \cdots$

3.
$$CH_3-\overset{\overset{\displaystyle O}{\|}}{C}-COO^- + {}^-OOC-CH_2-\overset{\overset{\displaystyle +NH_3}{|}}{CH}-COO^- \rightleftharpoons CH_3-\overset{\overset{\displaystyle +NH_3}{|}}{CH}-COO^- + \cdots$$

4.
$$HN=\overset{\overset{\displaystyle NH_2}{|}}{\underset{\underset{\displaystyle H}{|}}{C}}-N-CH_2-CH_2-CH_2-\overset{\overset{\displaystyle +NH_3}{|}}{CH}-COO^- + H_2O \longrightarrow H_2N-\overset{\overset{\displaystyle O}{\|}}{C}-NH_2 + \cdots$$

5.
$$H_2N-CH_2-CH_2-CH_2-\overset{\overset{\displaystyle +NH_3}{|}}{CH}-COO^- + H_2N-\overset{\overset{\displaystyle O}{\|}}{C}-O-\textcircled{P} \longrightarrow P_i + \cdots$$

Problem 6 Where are the two nitrogen atoms of urea derived from in the urea cycle?

A. γ-Amino group of ornithine
B. Carbamoyl phosphate nitrogen
C. α-Amino group of ornithine
D. Aspartate nitrogen

Catabolism of the Carbon Skeletons

Glucogenic, or glycogenic, amino acids, by definition, are potentially convertible to glucose. An amino acid is glucogenic if the catabolism of its carbon skeleton yields pyruvate, 3-phosphoglycerate, or a TCA-cycle intermediate, because these precursors can enter gluconeogenesis to produce glucose.

Ketogenic amino acids, by definition, yield acetyl-CoA or acetoacetyl-CoA on catabolism, both of which can be converted to ketone bodies. Leucine is the only purely ketogenic amino acid.

Five amino acids are both glucogenic and ketogenic, while the remainder are strictly glucogenic.

Learning all the intricate catabolic pathways for all the individual amino acids is a job for a biochemist specializing in amino-acid metabolism; medical students need not attempt such a colossal task. Instead, one need only focus on the general features of the medically relevant pathways outlined in this chapter.

AMINO ACID BIOSYNTHESIS

Bacteria and higher plants can synthesize all twenty common α-amino acids by adding ammonia to carbon skeletons, such as the α-keto acids, that they fabricate. Mammals, however, cannot manufacture certain α-keto acids that are required in the synthesis of their corresponding amino acid. Any amino acid that humans either cannot synthesize or are unable to manufacture in adequate quantity is termed *essential*.

An essential amino acid must be provided in the diet. The absence of an essential amino acid from the diet generally, but not always, causes a negative *nitrogen balance*; that is, the total nitrogen losses in the urine, feces, and sweat exceed the dietary nitrogen intake. Replacing the deficient essential amino acid promptly restores the positive nitrogen balance that is needed for growth or the zero nitrogen balance that is required for maintenance.

The absence of a nonessential amino acid from the diet, however, does not impair protein synthesis, because such an amino acid can be manufactured in adequate amounts. Essential amino acids are *not* more important in human metabolism than the nonessential amino acids; they differ only in that the essential amino acids must be provided in the diet. Humans could synthesize many, but not all, essential amino acids by transaminating the corresponding α-keto acids, but, since these keto acids cannot be synthesized and do not appear in the diet, these pathways are normally of no avail.

Ten amino acids are essential for humans: arginine, histidine, isoleucine, leucine, lysine, methionine, phenylalanine, threonine, tryptophan, and valine. Arginine and histidine differ from the other eight essential amino acids in that their short-term deficiency in humans does *not* produce a negative nitrogen balance. Nevertheless, humans cannot synthesize any histidine. Although man can synthesize arginine using the urea-cycle enzymes, most of this arginine is degraded to ornithine plus urea. Hence, man cannot synthesize enough arginine to meet his needs.

Cysteine and tyrosine are in the gray zone between essential and nonessential. The absence of cysteine from the diet, for example, raises the methionine requirement by 30%, because cysteine is synthesized from methionine. Similarly, the absence of tyrosine increases the phenylalanine requirement. Thus, tyrosine and cysteine are set apart from the other nonessential amino acids. The combined dietary methionine-cysteine and phenylalanine-tyrosine intakes are used to compute the requirements for these amino acids.

The essential amino acids can be classified as follows. An asterisk is used to denote the two amino acids that are nonessential but are synthesized from essential amino acids:

1. All three branched-chain amino acids (isoleucine, leucine, valine).
2. All three aromatic amino acids (phenylalanine-tyrosine,* tryptophan).
3. Both sulfur-containing amino acids (methionine-cysteine*).
4. All three basic amino acids (arginine, histidine, lysine).
5. Threonine.

The syntheses of essential amino acids by plants and bacteria are generally longer and more intricate than those for nonessential amino acids.

Each essential amino acid is required in differing amounts. Humans must ingest four times more leucine than tryptophan, for example. A protein of high biologic quality has a relative amino-acid composition that is similar to the human requirement. The ideal food protein is egg albumin. Other high quality protein sources include eggs, meats, fish, and poultry. Milk has moderately high quality protein. Plant proteins, such as those in corn and wheat, are generally low quality. Corn, for instance, has too little tryptophan and lysine. For every gram of egg albumin in the human diet, two or more grams of corn protein would be required to equal the nitrogen-retention provided by egg albumin, the nutritionally ideal protein. Of the amino acids in egg albumin, 40% are essential. Artificial proteins composed entirely of essential amino acids actually have a lower biologic quality than egg albumin, because they overtax the body's ability to synthesize the nonessential amino acids.

Problem 7	The α-amino groups derived from amino-acid catabolism are collected and stored as:

A. Urea
B. Aspartate

C. Glutamate
D. Carbamoyl phosphate

Problem 8	Which class of amino acids contains only nonessential amino acids?

A. Aromatic
B. Branched-chain
C. Basic

D. Sulfur-containing
E. Acidic

Problem 9	In calculating the biologic quality of a dietary protein, the combined cysteine-methionine content is more useful than the methionine content alone, because:

A. Humans can synthesize cysteine from methionine and vice versa.
B. Dietary cysteine spares methionine.
C. Both are essential amino acids.

Phenylalanine and Tyrosine Metabolism

As shown in Figure 13-3, phenylalanine hydroxylase transforms the essential amino acid phenylalanine into *p*-hydroxyphenylalanine, or tyrosine. Phenylalanine hydroxylase uses tetrahydrobiopterin, a pteridine compound resembling folic acid, as a coenzyme to reduce $NADP^+$ to NADPH.

The inherited absence of phenylalanine hydroxylase diverts phenylalanine metabolism into a minor pathway that is not employed normally. Figure 13-3 shows that in this pathway, the alanine portion of the molecule is transaminated to pyruvate, generating phenylpyruvate. This phenylketone spills into the urine; hence the name *phenylketonuria* (PKU) for the disorder resulting from the absence of this enzyme. Phenylpyruvate is reduced to phenyllactate and also decarboxylated to phenylacetate. These aromatic compounds damage the developing brain, producing mental retardation. The brain will develop normally, however, if the phenylalanine intake in the diet is extremely low. An artificial protein that lacks phenylalanine is given to children with PKU to maintain low serum levels of phenylpyruvate and its derivatives. Since they cannot synthesize tyrosine, this becomes an essential amino acid for those with PKU. Whether or not adult phenylketonurics may return to a normal diet without adverse effects is controversial. Certainly, an adult's brain is not as readily damaged by phenylketones as is the infant's developing brain.

In the normal pathway of tyrosine metabolism, tyrosine is transaminated to *p*-hydroxyphenylpyruvate, which in turn is oxidized to homogentisate by using ascorbate as a coenzyme. This is a typical example of the role of ascorbate in many oxygenation reactions. Homogentisate oxidase then opens the phenyl

Figure 13-3

Metabolism of phenylalanine and tyrosine.

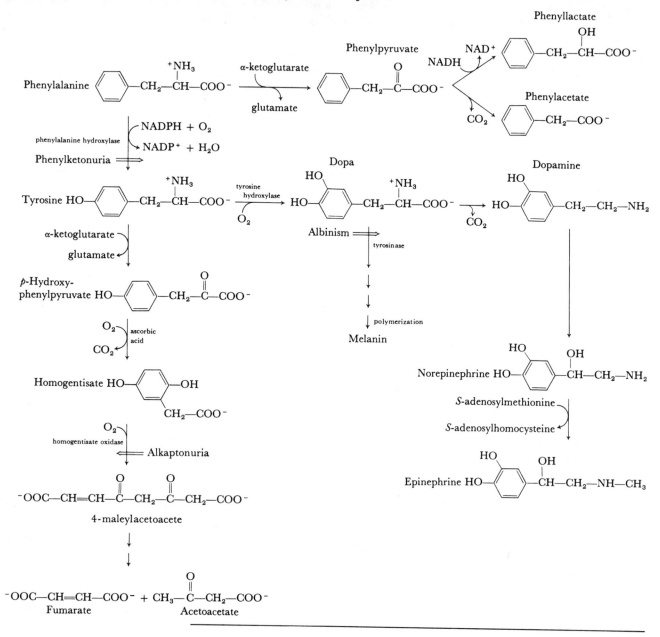

ring of homogentisate to create a straight-chain compound, which is then cleaved to generate fumarate and acetoacetate. Fumarate enters the TCA cycle, thereby rendering phenylalanine and tyrosine glucogenic. They are also ketogenic, because their catabolism yields acetoacetate.

The inherited deficiency of homogentisate oxidase, termed *alkaptonuria*, leads to the buildup of homogentisate derivatives that polymerize into melanin-like pigments. These pigments with stain connective tissue brownish black, and the urine from alkaptonuric patients darkens after standing due to pigment formation.

An alternative route in the normal pathway of tyrosine metabolism is its hydroxylation by tyrosine hydroxylase to *dihydroxyphenylalanine* (dopa). This amino acid can then be decarboxylated to dopamine, which is hydroxylated to the neurotransmitter, norepinephrine. *S*-Adenosylmethionine then donates its methyl group to norepinephrine to produce the sympathetic agent, epinephrine. Dopa is administered to patients with Parkinson's disease to correct their lack of dopamine in the basal ganglia of the brain. Another fate of dopa is its hydroxylation by tyrosinase and eventual polymerization to the pigment, melanin. The inherited deficiency of tyrosinase blocks melanin synthesis, which results in albinism.

Tyrosine is iodinated in the thyroid to mono- and diiodotyrosine. Two molecules of diiodotyrosine are joined to create thyroxine (tetraiodothyronine or T_4) and serine. Triiodothyronine (T_3), the most active thyroid hormone, is synthesized by adding mono- to diiodotyrosine or by deiodinating T4 to T3.

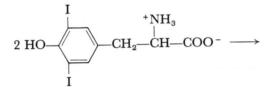

Diiodotyrosine

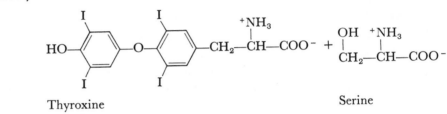

Thyroxine   Serine

Tryptophan Metabolism

Humans use tryptophan, an essential amino acid, to synthesize part of their nicotinic-acid requirement. Thus, a high tryptophan content in the diet can compensate for a low content of this vitamin. Pellagra classically strikes people subsisting on corn, because corn has little niacin or tryptophan.

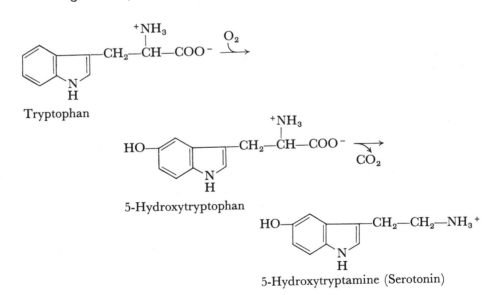

Tryptophan

5-Hydroxytryptophan

5-Hydroxytryptamine (Serotonin)

Tryptophan is also the precursor of 5-hydroxytryptamine, or serotonin, a neurotransmitter. Tryptophan must be hydroxylated to 5-hydroxytryptophan, which is decarboxylated to 5-hydroxytryptamine. Serotonin is then deaminated and oxidized to 5-hydroxyindoleacetic acid (5-HIAA). The urinary 5-HIAA excretion is elevated in the carcinoid syndrome.

Problems 10–14

Match the unbalanced reactions below to the descriptions in Problems 10–14:

A. Phenylalanine → phenylpyruvate
B. Phenylalanine → *p*-hydroxyphenylalanine
C. Dopa → dopamine
D. Norepinephrine → epinephrine

10. The principal route of normal phenylalanine metabolism
11. The alternative pathway of phenylalanine degradation in phenylketonuria
12. Transamination
13. Decarboxylation
14. Methylation

Branched-Chain Amino-Acid Catabolism

The three branched-chain amino acids—isoleucine, leucine, and valine—are all essential. They are degraded by transamination to α-keto acids, which are oxidized and decarboxylated to acyl-CoA.

The inherited absence of branched-chain keto-acid dehydrogenase causes these three α-keto acids to accumulate and spill into the urine. Their sweet smell conferred the name of "maple-syrup urine disease" to this ketoaciduria. The disorder is fatal unless the infants afflicted by it are fed artificial proteins that are low in branched-chain amino acids.

Valine and isoleucine catabolism eventually produces propionyl-CoA, a three-carbon, fatty-acyl-CoA. This is carboxylated to methylmalonyl-CoA. A mutase employing vitamin B_{12} then converts this four-carbon compound to succinyl-CoA.

Methionine and Cysteine Metabolism

Methionine, an essential amino acid, is the precursor of cysteine (see p. 166).

As shown in Figure 13-4, adenosine from ATP joins via the sulfur atom of methionine to generate *S*-adenosylmethionine. This important methyl donor is then able to liberate its sulfur-bound methyl group in reactions such as the methylation of phosphatidylethanolamine to phosphatidylcholine or the methylation of norepinephrine to epinephrine. The *S*-adenosylhomocysteine generated in these reactions then liberates its adenosine group to yield homocysteine, which differs from cysteine only in having an additional CH_2 group in its side chain. Cystathionine synthase then condenses homocysteine with serine to forming cystathionine. (Although cystathionine contains two amino acids, it is not a dipeptide, because they are not joined by a peptide linkage.) Cystathionine lyase next cleaves α-ketobutyrate (originally from homocysteine) from the sulfur atom, leaving this sulfur attached to the former serine molecule, which now becomes cysteine.

Figure 13-4

Conversion of methionine to cysteine.

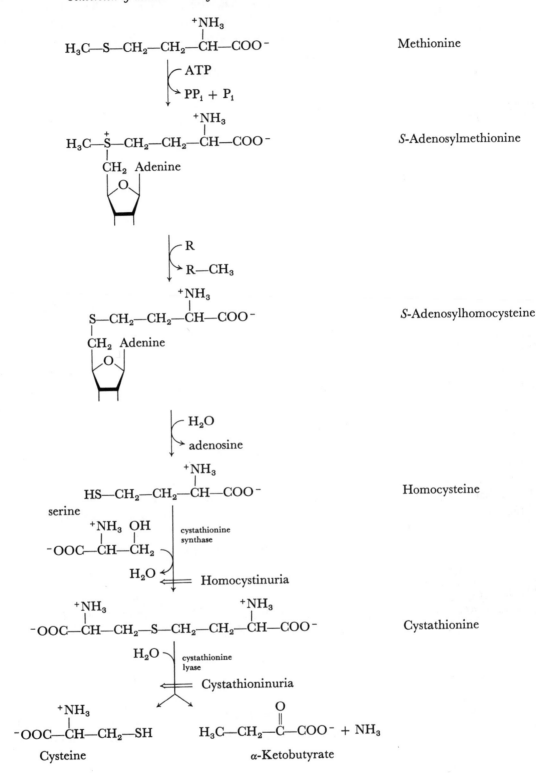

Homocystinuria, which results from the absence of cystathionine synthase, and cystathioninuria, which results from the deficiency of cystathionine lyase, both block cysteine synthesis, thus making it an essential amino acid in the strict sense. Both disorders respond to dietary methionine restriction and cysteine supplementation.

In further catabolism, cysteine degradation liberates sulfate.

Problems 15–18

Match the hereditary disorders below with the compounds that accumulate in the urine:

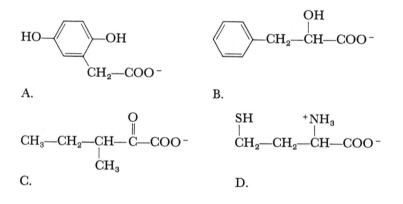

A.

B.

C.

D.

15. Maple-syrup urine disease
16. Homocystinuria
17. Alkaptonuria
18. Phenylketonuria

Basic Amino-Acid Metabolism

The three basic amino acids—arginine, histidine, and lysine—are all essential. The urea cycle supplies only part of the daily arginine requirement.

Histidine decarboxylase transforms histidine into histamine, a mediator of the inflammatory response in allergic reactions.

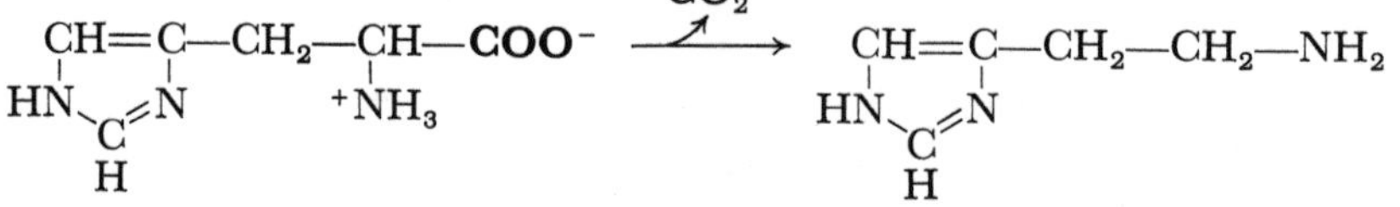

$$CH=C-CH_2-CH-COO^- \quad \xrightarrow{CO_2} \quad CH=C-CH_2-CH_2-NH_2$$

Histidine

Histamine

Hydroxylysine is synthesized from collagen-bound lysine in a manner similar to hydroxyproline formation (see next section on proline).

Glutamate, Glutamine, and Proline Metabolism

These nonessential amino acids all stem from α-ketoglutarate, which is transaminated to create glutamate. Glutamine synthetase adds NH_3 to glutamate to produce glutamine, which is actually an amide:

$$
\begin{array}{c}
\text{COO}^- \\
|\\
\text{CH}_2 \\
|\\
\text{CH}_2 \\
|\\
\text{CH}-{}^+\text{NH}_3 \\
|\\
\text{COO}^-
\end{array}
\quad + \text{NH}_3 + \text{ATP}
\;\rightleftharpoons\;
\begin{array}{c}
\text{O}\\
\|\\
\text{C}-\text{NH}_2 \\
|\\
\text{CH}_2 \\
|\\
\text{CH}_2 \\
|\\
\text{CH}-{}^+\text{NH}_3 \\
|\\
\text{COO}^-
\end{array}
\quad + \text{ADP} + \text{P}_i + \text{H}_2\text{O}
$$

Glutamate   Glutamine

The first step in proline synthesis, shown in Figure 13-5, is the reduction of the γ-carboxyl group of glutamate to an aldehyde, creating glutamate semialdehyde, which spontaneously becomes a cyclic compound. Further reduction then yields the imino acid, proline.

Only after incorporation into collagen can proline be oxygenated to hydroxyproline, using O_2 with ascorbate as a coenzyme. Coupled to this reaction is the oxidation and decarboxylation of α-ketoglutarate to succinyl-CoA.

Glutamate, glutamine, and proline are all degraded to α-ketoglutarate, a TCA-cycle intermediate; hence they are glucogenic.

Figure 13-5   *Synthesis of proline and hydroxyproline from glutamate.*

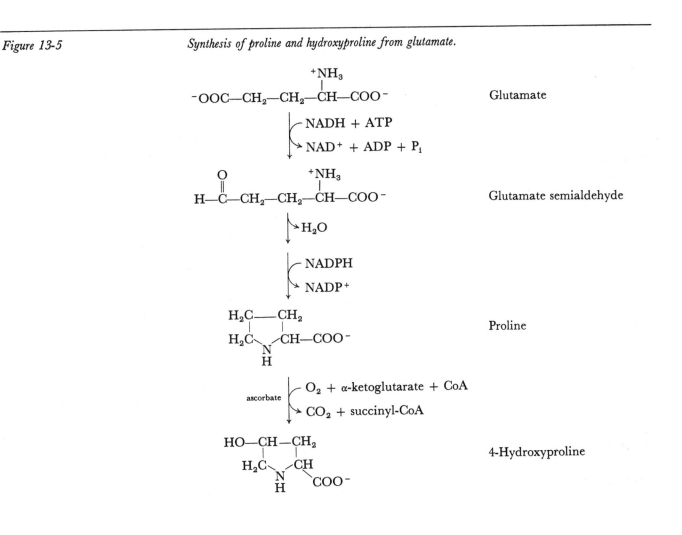

Aspartate, Asparagine, and Alanine Metabolism

Aspartate and alanine arise from the transaminations of oxaloacetate and pyruvate, respectively:

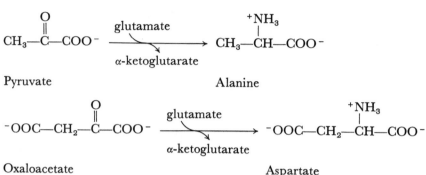

Asparagine synthetase amidates aspartate to produce asparagine, the β-amide of aspartic acid. Conversely, asparaginase deamidates asparagine to aspartate.

Asparaginase is injected into children with acute lymphocytic leukemia to induce remission of the disorder by depriving leukemic cells of asparagine; normal cells are less sensitive to asparagine deprivation.

Glycine, Serine, and Threonine Metabolism

3-Phosphoglycerate from glycolysis is the precursor to serine.

Tetrahydrofolic acid (THFA) interconverts serine and glycine by removal or addition of a methylene group from N^5, N^{10}-methylene-THFA:

$$\underset{\text{Serine}}{\overset{\text{OH}\quad{}^+\text{NH}_3}{\text{CH}_2-\text{CH}-\text{COO}^-}} + \text{THFA} \rightleftharpoons \underset{\text{Glycine}}{\overset{{}^+\text{NH}_3}{\text{CH}_2-\text{COO}^-}} + \text{CH}_2\text{-THFA} + \text{H}_2\text{O}$$

Threonine, an essential amino acid, is degraded to yield glycine and acetyl-CoA. Since glycine can be converted to 3-phosphoglycerate, threonine is glucogenic. The acetyl-CoA produced makes threonine also ketogenic, since this acetyl-CoA can be converted to ketone bodies.

HORMONAL CONTROL OF AMINO ACID METABOLISM

Insulin promotes the active transport of amino acids across cell membranes. It inhibits gluconeogenesis from amino acids, and it promotes protein synthesis. Growth hormone (STH), androgens, and thyroid hormones also promote a positive nitrogen balance by stimulating protein synthesis.

Glucocorticoids, on the other hand, enhance gluconeogenesis from amino acids. The resultant negative nitrogen balance in high-dose chronic glucocorticoid therapy leads to thinning of the skin and osteoporosis.

Following a meal, insulin release fosters tissue uptake of amino acids and subsequent protein synthesis. During fasting, however, the glucocorticoids predominate, and insulin virtually disappears. The protein in skeletal muscle and, eventually, in the viscera is eroded daily by gluconeogenesis to provide glucose. Humans have no amino-acid storage capacity as such; useful tissue proteins must be sacrificed to supply glucose for the brain.

Dietary carbohydrate is protein-sparing. The presence of at least 100 g of dietary carbohydrate daily for adults prevents excess protein catabolism via gluconeogenesis. Much of the protein in a carbohydrate-deficient diet is not used for protein synthesis; instead, it enters gluconeogenesis to compensate for a lack of glucose. Hence, the nitrogen balance is usually negative.

Problem 19

Which process is *not* part of thyroxine synthesis in the thyroid gland?

A. iodination of tyrosine
B. formation of a diiodotyrosine dipeptide
C. liberation of serine as two molecules of diiodotyrosine join

---

Problem 20

Histidine is converted to histamine by:

A. transamination
B. hydroxylation
C. decarboxylation
D. reduction with NADH

Problem 21

Carcinoid tumors in the liver rapidly metabolize which amino acid to produce excess urinary 5-hydroxyindoleacetic acid (5-HIAA)?

A. isoleucine
B. hydroxyproline
C. alanine
D. tryptophan

Problem 22

The synthesis of hydroxyproline from proline requires all of the following *except*:

A. free proline
B. ascorbate
C. CoA
D. α-ketoglutarate
E. O_2

Problem 23

The "Dr. Atkin's diet" severely restricts carbohydrate but allows ample fat and protein intake. Its metabolic consequences are:

A. protein-sparing
B. relatively normal metabolism
C. ketosis with protein-wasting
D. positive nitrogen balance

Problem 24

The methylene group transferred to glycine in converting it to serine comes from:

A. S-adenosylmethionine
B. N^5,N^{10}-methylene-THFA
C. methylene-B_{12}
D. carboxybiotin

A high intake of which amino acid can prevent pellagra in a niacin-deficient diet?

A. lysine   C. threonine
B. methionine   D. tryptophan

In treating Parkinson's disease with the amino acid dopa, you should tell your patients not to take multivitamins containing the one vitamin that would enhance dopa metabolism. This vitamin is:

A. pyridoxal phosphate   C. biotin
B. thiamine   D. niacin

ANSWERS

1. $^-OOC-CH_2-CH_2-CH-COO^-$
   with $^+NH_3$ on the CH

   Glutamine is hydrolyzed by glutaminase to glutamate and NH_3.

2. $H_2N-\overset{\overset{\displaystyle O}{\|}}{C}-O-\text{(P)}$

   Carbamoyl-phosphate synthetase condenses carbon dioxide, ammonia, and water to generate carbamoyl phosphate for the urea cycle.

3. $^-OOC-CH_2-\overset{\overset{\displaystyle O}{\|}}{C}-COO^-$

   Pyruvate and aspartate are transaminated to yield alanine and oxaloacetate. Several transaminases, using pyridoxal phosphate as a coenzyme, can catalyze this reaction.

4. $H_2N-CH_2-CH_2-CH_2-\overset{\overset{\displaystyle +NH_3}{|}}{CH}-COO^-$

   Arginase in the cytoplasm cleaves arginine to urea and ornithine.

5. $H_2N-\overset{\overset{\displaystyle O}{\|}}{C}-NH-CH_2-CH_2-CH_2-\overset{\overset{\displaystyle +NH_3}{|}}{CH}-COO^-$

   Ornithine carbamoyltransferase in the mitochondria condenses ornithine with carbamoyl phosphate to yield citrulline and P_i.

6. B, D.
7. C.
8. E. (Glutamate and aspartate, the acidic amino acids, are both nonessential.)
9. B. (Humans synthesize cysteine from the essential amino acid methionine, but they cannot synthesize methionine from cysteine.)
10. B. (*p*-Hydroxyphenylalanine is tyrosine.)
11. A.
12. A. (The alanine portion of phenylalanine is transaminated with α-ketoglutarate to form pyruvate.)
13. C.
14. D.

15. C. (This branched chain α-keto acid cannot be degraded without the branched-chain keto-acid dehydrogenase.)
16. D. (This is homocysteine.)
17. A. (Homogentisate accumulates in homogentisate-oxidase deficiency.)
18. B. (This is phenyllactate, which, along with phenylpyruvate, is produced in phenylketonuria.)
19. B. (There is no peptide bond between the two tyrosine molecules.)
20. C.
21. D. (The indole ring in 5-HIAA comes from tryptophan.)
22. A. (The proline must be bound in a protein such as collagen; free proline cannot be hydroxylated.)
23. C.
24. B.
25. D. (Humans synthesize part of their niacin requirement from tryptophan.)
26. A. (Pyridoxal phosphate is an essential coenzyme for dopa decarboxylase, which converts dopa to dopamine. Giving supplemental pyridoxal phosphate will promote this conversion in many organs and will lower the blood level of dopa, thereby reducing the amount of dopa that crosses the blood-brain barrier to correct the biochemical imbalance in Parkinson's disease; (dopamine does not cross this barrier. Thus, dopa therapy will become less effective.)

REFERENCES

Fischer, J. E., Yoshimura, N., Aguirre, A., James, J. H., Cummings, M. G., Abel, R. M., and Deindoerfer, F. Plasma amino acids in patients with hepatic encephalopathy. Effects of amino acid infusions. *Am. J. Surg.* 127:40, 1974.

Garza, C., Scrimshaw, N. S., and Young, V. R. Human protein requirements: a long-term metabolic nitrogen balance study in young men to evaluate the 1973 FAO/WHO safe level of egg protein intake. *J. Nutr.* 107:335, 1977.

Goodhart, R. S., and Shils, M. E. *Modern Nutrition in Health and Disease* (5th ed.). Philadelphia: Lea & Febiger, 1973. Pp. 28–88.

Goodman, L. S., and Gilman, A. *The Pharmacological Bases of Therapeutics* (5th ed.). New York: Macmillan, 1975. Pp. 1482 (glucocorticoids), 1514–1516 (insulin).

Lehninger, A. L. *Biochemistry: The Molecular Basis of Cell Structure and Function* (2nd ed.). New York: Worth, 1975. Pp. 559–586, 693–699.

Merino, G. E., Jetzer, T., Doizaki, W., and Najarian, J. S. Methionine-induced hepatic coma in dogs. *Am. J. Surg.* 130:41, 1975.

Simpson, D. P. Control of hydrogen ion homeostasis and renal acidosis. *Medicine* 50:503, 1971.

White, A., Handler, P., and Smith, E. L. *Principles of Biochemistry* (5th ed.). New York: McGraw-Hill, 1973. Pp. 629–655, 677–702.

Nucleoside Metabolism and Biosynthesis

The digestion of dietary nucleoproteins begins with their proteolysis in the stomach and intestine. Once liberated from their protein coating, the polynucleotides are cleaved by pancreatic ribonucleases and deoxyribonucleases to oligo- and mononucleotides. After absorption, most dietary nucleotides are catabolized rather than incorporated into human nucleic acids. Man does not require any dietary purines or pyrimidines.

PYRIMIDINE BIOSYNTHESIS

Required in both pyrimidine and purine biosynthesis is 5-phosphoribosylpyrophosphate (PRPP), which is generated by pyrophosphorylating ribose-5-P with ATP. The hexose-monophosphate shunt provides ribose-5-P. The enzyme that catalyzes this pyrophosphorylation, ribose-phosphate pyrophosphokinase, is allosterically inhibited by ADP and GDP.

$$\text{Ribose-5-P} + \text{ATP} \longrightarrow \text{PRPP} + \text{AMP}$$

Pyrimidine synthesis also requires carbamoyl phosphate. Carbamoyl-phosphate synthase (glutamine) in the cytoplasm deaminates glutamine and combines the NH_3 obtained with CO_2 and H_2O to yield carbamoyl phosphate. The urea cycle in the mitochondria utilizes another enzyme, carbamoyl-phosphate synthetase (ammonia), which does not require glutamine (see Fig. 13-2). This difference between these two enzymes is indicated by appending the name of the substrate in parentheses.

$$\text{Glutamine} + \text{CO}_2 + \text{H}_2\text{O} + 2\text{ATP} \longrightarrow$$
$$\text{glutamate} + \text{carbamoyl phosphate} + 2\text{ADP} + \text{P}_i$$

Aspartate transcarbamoylase, the rate-controlling enzyme in pyrimidine synthesis, links the carbamoyl group of the unstable carbamoyl phosphate to aspartate to generate *N*-carbamoylaspartate, as shown in Figure 14-1. CTP is an inhibitory modifier for this allosteric enzyme. Since *N*-carbamoylaspartate is used only in pyrimidine biosynthesis, this is the committed step of the pathway.

Dehydration transforms the straight-chain *N*-carbamoylaspartate into a cyclic pyrimidine, dihydroorotic acid, which is oxidized with NAD^+ to orotic acid. Next, orotate phosphoribosyltransferase adds PRPP to orotic acid to produce orotidine-5-P, a nucleoside phosphate or nucleotide. Orotidine-5-P decarboxylase then removes the carboxyl group from the side chain to create uridylic acid or UMP.

The rare, hereditary absence of orotate phosphoribosyltransferase and orotidine-5-P decarboxylase blocks UMP synthesis and leads to orotic aciduria.

Figure 14-1   *Pyrimidine biosynthesis.*

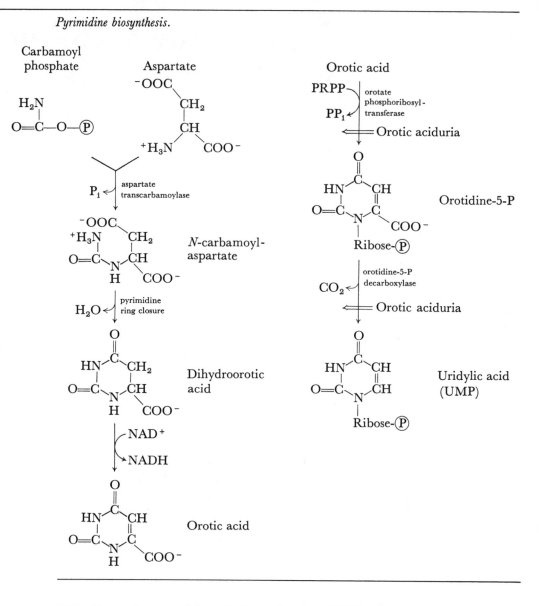

This disease is treated by administering large doses of uridine to bypass this enzymatic block and supply a precursor for synthesizing UTP, CTP, and TTP. This CTP will then allosterically inhibit aspartate transcarbamoylase, thereby reducing orotic-acid production.

UMP is the precursor of all the pyrimidine nucleotides. ATP phosphorylates UMP first to UDP and then to UTP:

$$\text{UMP} \xrightarrow[\text{ATP} \quad \text{ADP}]{} \text{UDP} \xrightarrow[\text{ATP} \quad \text{ADP}]{} \text{UTP}$$

The uracil of UTP is aminated with glutamine to create the cytosine in CTP:

180 14. Nucleoside Metabolism and Biosynthesis

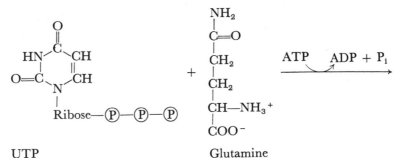

UTP   Glutamine

CTP   Glutamate

5-Iodo-2′-deoxyuridine is a pyrimidine analog used clinically to treat keratitis due to herpes virus. This iodinated pyrimidine becomes incorporated into viral DNA during DNA replication and interferes with DNA function.

SYNTHESIS OF DEOXYRIBO-NUCLEOTIDES

In humans, all the nucleotides are synthesized initially as ribonucleotides. Once this synthesis is complete, the ribose in the nucleotide is reduced. The mechanism for this reduction in humans is unclear, but it may resemble the thioredoxin system in *Escherichia coli*. NADPH reduces a disulfide bond in the protein thioredoxin to the free sulfhydryl groups. The hydrogen atoms carried on these sulfur atoms of thioredoxin then reduce the 2′ hydroxy group in ribonucleotides to create deoxyribonucleotides.

Thymidylate synthetase uses N^5,N^{10}-methylene-THFA to methylate dUMP to create dTMP and dihydrofolic acid (DHFA):

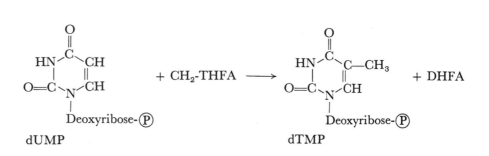

dUMP   dTMP

Folic acid antagonists, e.g., methotrexate, are used to treat leukemia because they block folate-dependent steps in nucleic-acid synthesis, such as the thymidylate-synthetase reaction. Blocking this crucial step in DNA replication slows the proliferation of both leukemic cells and those of normal bone marrow; the latter effect causes macrocytic anemia similar to that seen in folate deficiency.

5-Fluorouracil (5-FU) is used to treat many solid tumors in humans. Once activated to 5-fluoro-dUMP, it competitively inhibits thymidylate synthetase, because it resembles dUMP.

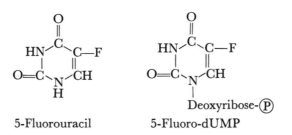

5-Fluorouracil   5-Fluoro-dUMP

Cytosine arabinoside, like 5-fluorouracil, is a pyrimidine analog. Its medical use lies in treating acute leukemias. Although its exact mechanism of action is uncertain, the arabinose sugar that replaces the normal deoxyribose interferes with base stacking when this analog is incorporated into DNA.

Pyrimidine catabolism yields β-alanine, which may be incorporated into certain human dipeptides or further degraded to CO_2. The pyrimidine nitrogens are liberated as ammonia, which can enter the urea cycle.

Problems 1–4

Match the enzymes below to the descriptions in Problems 1–4:

A. orotidine-5-P decarboxylase
B. aspartate transcarbamoylase
C. carbamoyl-phosphate synthase (glutamine)
D. carbamoyl-phosphate synthetase (ammonia)
E. orotate phosphoribosyltransferase

1. Inherited dual enzyme-deficiency disease causing inadequate pyrimidine synthesis.
2. Supplies nitrogen to the urea cycle.
3. Rate-controlling step of pyrimidine biosynthesis.
4. Generates the unstable carbamoyl phosphate for pyrimidine biosynthesis.

Problem 5

Which statement about aspartate transcarbamoylase is *incorrect*?

A. It obeys Michaelis-Menten enzyme kinetics.
B. Its allosteric inhibitor is CTP.
C. It catalyzes the committed step of pyrimidine biosynthesis.
D. It is a multiple-subunit enzyme.

Match the drugs below to the statements in Problems 6–8:

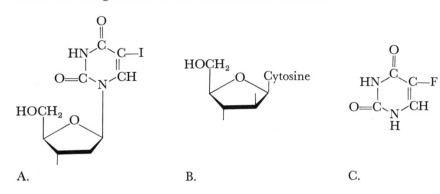

A.   B.   C.

6. Arabinose interferes with base stacking after this nucleoside is incorporated into polynucleotides.
7. Used to treat keratitis due to herpes virus.
8. Analog of dUMP that inhibits thymidylate synthetase.

Problem 9

Where do the nitrogen atoms of the pyrimidine ring originate during pyrimidine biosynthesis?

A. Glutamine and NH_3
B. Carbamoyl phosphate and glutamate
C. Aspartate and carbamoyl phosphate
D. Glutamine and glutamate

PURINE SYNTHESIS

Since purines contain both a pyrimidine and an imidazole ring, their synthesis is more complex than that of pyrimidines. In purine synthesis, ribose is bound to the growing purine ring from the outset rather than being attached after ring formation, as occurs in pyrimidine synthesis.

In the initial step of purine synthesis, shown in Figure 14-2, amidophosphoribosyl transferase (or amidotransferase) replaces the pyrophosphate group of PRPP with the amide amino group of glutamine, creating 5-phosphoribosylamine, an amino sugar phosphate. This amidotransferase reaction is the rate-controlling step of inosinic acid synthesis. Its allosteric inhibitors include GMP, GDP, GTP, AMP, ADP, and ATP.

In the next step, an entire glycine molecule is incorporated into the growing imidazole ring. The α-carboxyl group of glycine forms an amide with the amino group of 5-phosphoribosylamine to yield 5-phosphoribosylglycinamide. This addition of glycine is the only step in purine synthesis in which more than one member of the purine ring is added at a time.

N^5,N^{10}-Methenyltetrahydrofolate then adds its formyl group to the α-amino end of glycinamide to create 5-phosphoribosyl-N-formylglycinamide. This compound contains all five members of the imidazole ring, although no ring yet exists.

Next, another glutamine donates its amide amino group, which displaces the keto group, thereby transforming the N-formylglycinamide to N-formylglycinamidine. This amino group becomes the third member of the growing

Figure 14-2

Purine biosynthesis. (*Adapted from A. L. Lehninger, Biochemistry [2nd ed.]. New York: Worth, 1975. Fig. 26-2.*)

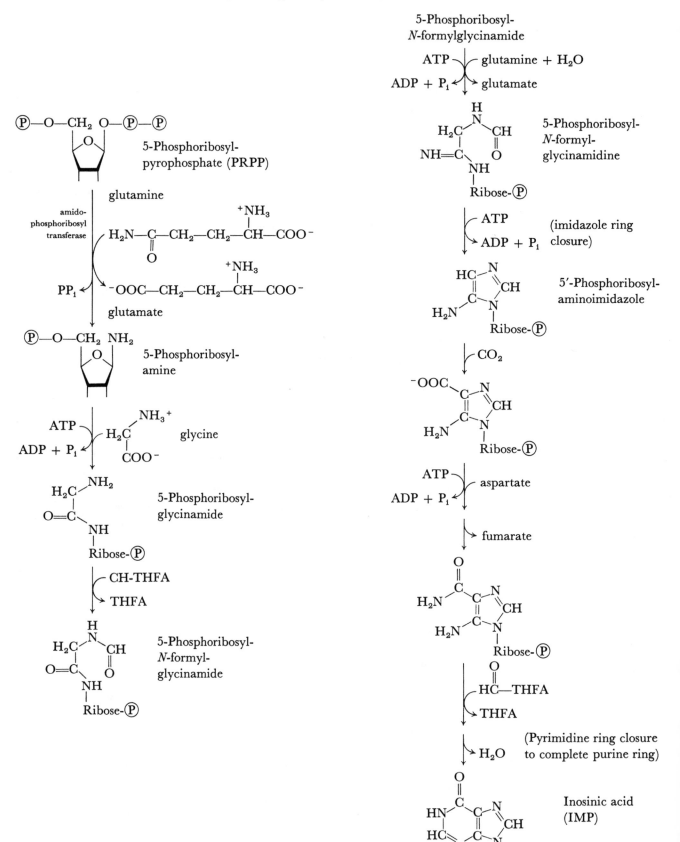

pyrimidine ring. Now the imidazole ring closes, leaving an aminoimidazole sugar phosphate.

The remaining reactions add the three additional members of the pyrimidine ring. First CO_2, then the α-amino group of aspartate, and finally a formyl group from tetrahydrofolate join to complete the purine ring. The first purine produced by this pathway is inosinic acid or IMP. The origin of its constituents is shown below:

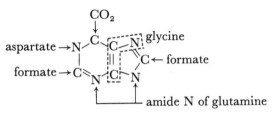

Figure 14-3 illustrates the pathway for synthesizing AMP and GMP from IMP. Oxidation of IMP with NAD^+ yields xanthylic acid, which is aminated

Figure 14-3   *Synthesis of adenylic and guanylic acids from inosinic acid. (Adapted from A. L. Lehninger, Biochemistry [2nd ed.]. New York: Worth, 1975. Figs. 26-5 and 26-6.)*

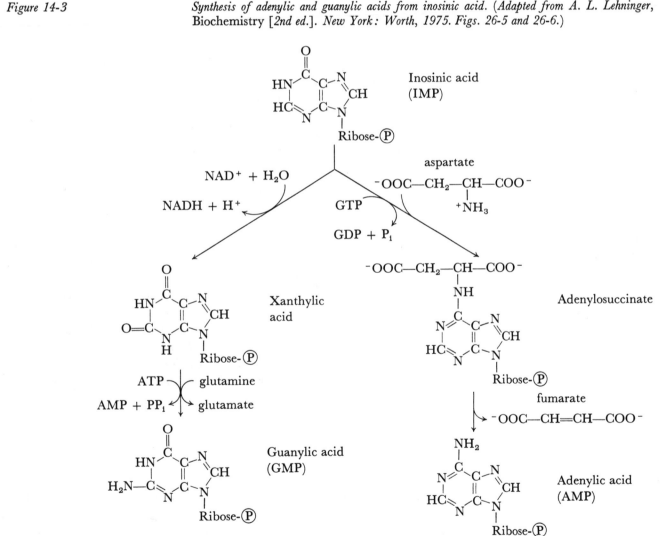

using glutamine to produce GMP. IMP may also bind to aspartate to form adenylosuccinate. Fumarate is then cleaved from adenylosuccinate, leaving AMP. In effect, aspartate aminates IMP to yield AMP.

GMP provides negative feedback to regulate its own formation by inhibiting the oxidation of IMP to xanthylic acid. Similarly, AMP inhibits the formation of its precursor, adenylosuccinate, from IMP. Folic acid antagonists block the two folate-dependent steps in IMP synthesis.

Two purine analogs used clinically to treat acute leukemias are 6-mercaptopurine and thioguanine, analogs of hypoxanthine and guanine, respectively. Both are activated in vivo after joining ribose phosphate. 6-Mercaptopurine bound to ribose phosphate inhibits the conversions of IMP to both xanthylic acid and adenylosuccinate, as well as inhibiting the amidophosphoribosyltransferase reaction.

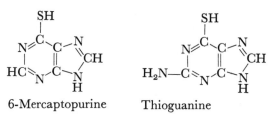

6-Mercaptopurine   Thioguanine

Nucleosidemonophosphate kinase phosphorylates any nucleoside monophosphate to its diphosphate form using ATP. Similarly, nucleosidediphosphate kinase transforms nucleoside diphosphates to triphosphates:

$$NMP + ATP \underset{\text{kinase}}{\overset{\text{nucleosidemonophosphate}}{\rightleftharpoons}} NDP + ADP$$

$$NDP + ATP \underset{\text{kinase}}{\overset{\text{nucleosidediphosphate}}{\rightleftharpoons}} NTP + ADP$$

PURINE CATABOLISM AND SALVAGE

AMP and GMP released from nucleic acids by the action of nucleases are hydrolyzed to yield the nucleosides adenosine and guanosine, as shown in Figure 14-4. The high-energy phosphate groups expended in phosphorylating the nucleosides are not regained in the reverse reactions. These nucleosides then liberate their ribose, leaving adenine and guanine.

Two salvage pathways exist to regenerate AMP and GMP from adenine and guanine, respectively. *Adenine phosphoribosyltransferase* adds ribose phosphate from PRPP to adenine to regenerate AMP. *Guanine (hypoxanthine) phosphoribosyltransferase* catalyzes the analogous reaction with guanine or hypoxanthine.

$$Adenine + PRPP \longrightarrow AMP + PP_i$$

$$Guanine + PRPP \longrightarrow GMP + PP_i$$

In Lesch-Nyhan syndrome, an inherited absence of guanine (hypoxanthine) phosphoribosyltransferase, excessive purine synthesis results in order to compensate for the lack of guanine salvage. The enhanced purine synthesis leads to

Figure 14-4

Pathways of purine salvage and catabolism to uric acid.

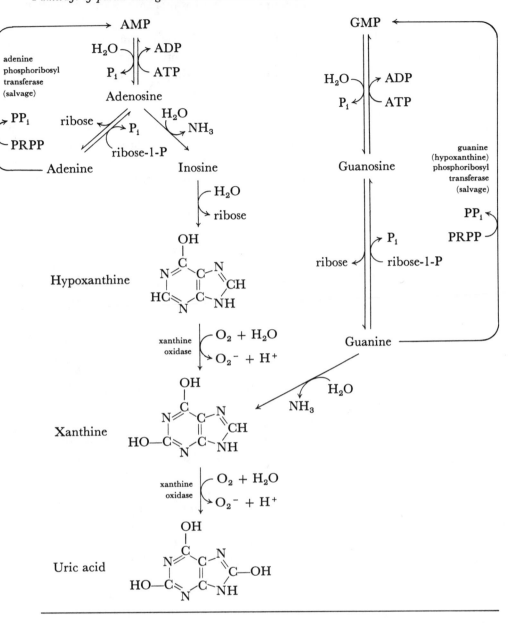

increased purine catabolism and consequent uric-acid production. Hyper-uricemia (high serum uric acid concentration) ensues, along with self-mutilation and mental retardation.

A second pathway for purine salvage adds ribose-1-P to adenine and guanine to create adenosine and guanosine, respectively, which are phosphorylated with ATP to yield AMP and GMP.

Besides being converted to adenine, adenosine may also be deaminated to inosine. Inosine then liberates its ribose, forming hypoxanthine. Xanthine oxidase, an enzyme that contains molybdenum and iron, adds O_2 and water to hypoxanthine to create xanthine and the superoxide radical O_2^-, which is converted to H_2O_2. (Catalase then splits hydrogen peroxide to yield H_2O +

$\frac{1}{2}O_2$.) Guanine, too, may be deaminated to produce xanthine. In the final step of human purine catabolism, xanthine oxidase transforms xanthine into uric acid.

Once it is synthesized, humans cannot degrade the purine ring itself. Thus, the purine derivative uric acid is the end product of human purine catabolism. The human kidney disposes of most of man's uric acid by glomerular filtration and tubular secretion. The renal tubules, however, reabsorb much of the filtered uric acid.

Approximately one quarter of man's daily uric acid removal occurs via intestinal bacteria, which possess urate oxidase to transform uric acid into allantoin. Other bacterial enzymes can degrade allantoin to allantoic acid and then to urea. Certain bacteria can then split urea to ammonia and carbon dioxide.

Gout, caused by hyperuricemia, is among the most common metabolic disorders in America, affecting about one in 200 adults. The clinical expressions of gout stem from the low solubility of sodium urate, which precipitates in joints (gouty arthritis), connective tissue (tophi), and the kidneys (urate stones). Hyperuricemia may be due to the overproduction of purines or the underexcretion of urate by the kidneys, or both. The overactivity of two enzymes in purine synthesis can cause human gout: ribose-phosphate pyrophosphokinase, which synthesizes PRPP, and amidophosphoribosyl transferase, which catalyzes the rate-limiting step in IMP synthesis.

An excess supply of ribose-5-P occurs in glucose-6-phosphatase deficiency (type-I glycogen storage disease) because the accumulating glucose-6-P is converted to ribose-5-P by the pentose-phosphate pathway. This, too, causes hyperuricemia.

Defective guanine salvage in Lesch-Nyhan syndrome also causes purine overproduction, and excessive purine synthesis occurs in diseases involving rapid cellular proliferation, such as leukemia.

The renal underexcretion of urate occurs in many kidney diseases. Substances that inhibit the tubular secretion of urate include the thiazide and "loop" diuretics, β-hydroxybutyrate (which abounds in starvation and diabetic ketoacidosis), and lactic acid (which is found in high levels in lactic acidosis and lactic acidemia, as well as in glucose-6-phosphatase deficiency).

Allopurinol is an analog to hypoxanthine that is used clinically to reduce the serum uric-acid level. It is not a true purine, because it lacks an imidazole ring (note the transposition of the C and N atoms in the formulas below). At low doses, it competitively inhibits xanthine oxidase, because the enzyme uses it as a substrate to synthesize alloxanthine. At higher doses, this alloxanthine derivative noncompetitively inhibits xanthine oxidase. After allopurinol is administered, the composition of the urinary purines excreted changes from predominantly

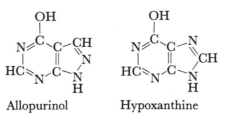

Allopurinol   Hypoxanthine

uric acid to a mixture of hypoxanthine, xanthine, and uric acid. Being far more soluble than uric acid, hypoxanthine and xanthine do not readily precipitate to form kidney stones.

Problem 10

Some patients develop gout because their amidophosphoribosyl transferase (or amidotransferase) enzyme is overactive. This enzyme:

A. synthesizes PRPP from ribose-5-P.
B. adds glycine to the growing pyrimidine ring.
C. aminates PRPP with glutamine.
D. aminates the pyrimidine ring of purines.

Problem 11

Pyrimidine catabolism:

A. produces uric acid.
B. is excessive in gout.
C. requires xanthine oxidase.
D. yields ammonia and carbon dioxide that can be converted to urea.

Problem 12

In what order are the three components of purine nucleotides produced or attached during synthesis?

A. pyrimidine ring, imidazole ring, then ribose-5-P.
B. ribose-5-P, pyrimidine ring, then imidazole ring.
C. imidazole ring, ribose-5-P, then pyrimidine ring.
D. ribose-5-P, imidazole ring, then pyrimidine ring.

Problem 13

During purine synthesis, one amino acid is entirely incorporated into the purine ring-structure. This amino acid, radioactively labeled, has been administered to humans, and the resultant labeled uric acid recovered to estimate their daily urate production. Which amino acid is used?

A. aspartate   C. glutamine
B. glycine   D. glutamate

Problem 14

Folic acid deficiency produces megaloblastic, macrocytic anemia by slowing the folate-dependent:

A. steps in pyrimidine synthesis.
B. conversion of IMP to AMP and GMP.
C. steps in purine synthesis and the conversion of dUMP to dTMP.
D. reduction of ribose to deoxyribose.

| Problem 15 | Below are five causes of hyperuricemia, or gout. Which one involves the over-production of uric acid, rather than its renal underexcretion? |

A. starvation   D. chronic renal failure
B. leukemia   E. alcoholic ketoacidosis
C. lactic acidosis

| Problem 16 | Which rare enzyme deficiency leads to hypouricemia (low serum uric-acid)? |

A. xanthine oxidase
B. carbamoyl-phosphate synthase (glutamine)
C. glutamate dehydrogenase
D. orotidine-5-P decarboxylase

ANSWERS

1. A, E. (The disease is orotic aciduria.)
2. D. (This mitochondrial enzyme utilizes NH_3 as its nitrogen source, and the carbamoyl phosphate produced enters the urea cycle.)
3. B.
4. C. (This cytoplasmic enzyme deaminates glutamine to supply NH_3 for carbamoyl-phosphate synthesis.)
5. A. (No allosteric enzyme obeys Michaelis-Menten kinetics.)
6. B. (Cytosine arabinoside.)
7. A. (5-Iodo-2'-deoxyuridine.)
8. C. (5-Fluorouracil.)
9. C.
10. C.
11. D. (Statements A, B, and C are true of *purine* catabolism.)
12. D.
13. B.
14. C.
15. B.
16. A.

REFERENCES

Bhagavan, N. V. *Biochemistry—A Comprehensive Review.* Philadelphia: Lippincott, 1974. Pp. 386–424.

Goodman, L. S., and Gilman, A. *The Pharmacological Basis of Therapeutics* (5th ed.). New York: Macmillan, 1975. Pp. 1268–1283 (methotrexate, pyrimidine, and purine analogs).

Lehninger, A. L. *Biochemistry: The Molecular Basis of Cell Structure and Function* (2nd ed.). New York: Worth, 1975. Pp. 729–747.

White, A., Handler, P., and Smith, E. L. *Principles of Biochemistry* (5th ed.). New York: McGraw-Hill, 1973. Pp. 705–733.

Wyngaarden, J., and Kelley, W. Gout. In Stanbury, J., Wyngaarden, J., and Fredrickson, D. (Eds.), *The Metabolic Basis of Inherited Disease* (3rd ed.). New York: McGraw-Hill, 1972.

15 DNA Replication, Transcription, and Translation

The most elucidating discoveries in molecular genetics have come from viral and bacterial studies. Although many of these breakthroughs apply to human genetics, some do not. Because human molecular genetics has not been adequately explored, this chapter will often mention bacterial and viral processes that may or may not occur in humans.

DNA REPLICATION

The double helix of DNA is replicated in a *semiconservative* manner; that is, each strand fathers a complementary strand so that the offspring DNA contain one parental strand and one newly synthesized strand. Human DNA, unlike that of *Escherichia coli*, is intertwined with basic proteins called *histones*, and it may require "unmasking" before DNA replication can begin.

DNA replication in *E. coli* and humans seems to share many common features. To begin the process in *E. coli*, the enzyme *DNA-directed RNA polymerase* binds to an initiating site on the DNA (RNA viruses use RNA-directed RNA polymerase to copy their RNA). *Unwinding proteins* then open a segment of the DNA double helix, allowing this DNA-directed RNA polymerase to synthesize two polyribonucleotide *priming chains* that are complementary to the DNA strands. Like all enzymes that synthesize polynucleotides, RNA polymerase cleaves pyrophosphate from nucleoside triphosphates (NTPs) as it adds NMPs to lengthen the chain in a 5′-to-3′ direction. This pyrophosphate is then hydrolyzed to phosphate, liberating energy to drive the reaction to the right:

$$NTP + \text{polynucleotide} \longrightarrow NMP\text{-polynucleotide} + PP_i$$
$$PP_i + H_2O \longrightarrow 2P_i$$

Next, a *DNA polymerase* adds deoxyribonucleotides to the 3′ end of this priming RNA, thereby replicating a short DNA segment termed an *Okazaki fragment*. *Endonucleases* then remove the RNA primers, and DNA polymerase fills in the gaps between the Okazaki fragments with deoxyribonucleotides. Finally, *DNA ligase* joins the ends of these fragments to create an entire DNA chain.

Note that DNA polymerases cannot initiate DNA replication. Instead, RNA polymerase begins this process by creating an RNA primer that is later excised.

DNA REPAIR

Agents that damage DNA include ultraviolet radiation, heat, extreme pH, and chemicals that alter purines and pyrimidines. One of the most frequent causes of DNA damage is ultraviolet radiation, which induces adjacent thymine rings to dimerize via covalent bonds. Thymine dimers block both DNA replication and transcription.

One method to correct thymine dimers, for example, uses an endonuclease to cut the DNA chain adjacent to the dimer. DNA polymerase then excises the dimer and replaces it with two normal thymine nucleotides. DNA ligase then reconnects this dinucleotide to the chain. A second method to correct thymine dimers uses an enzyme activated by visible light to cleave the covalent bonds that link the thymine residues together.

Xeroderma pigmentosum, a rare human photosensitivity disease, arises because of inherited endonuclease deficiency. Thymine dimers created in DNA molecules after solar irradiation, therefore, cannot be adequately repaired.

Alkylating agents are used to treat leukemias, lymphomas, and Hodgkin's disease. Their main mode of action is to alkylate the guanine in DNA, which leads to abnormal base-pairing of the altered guanine with thymine and guanine rather than with cytosine. This DNA damage blocks cell division in the tumor as well as in normal cells.

Bleomycin, another antineoplastic drug, acts by fragmenting DNA.

Several other rare disorders caused by faulty DNA repair include ataxia-telangiectasia, Bloom's syndrome, and Fanconi's anemia.

TRANSCRIPTION

Synthesis of complete RNA molecules from DNA is termed *transcription*. In DNA replication, the creation of the short RNA primers to initiate the process (see above) is not truly transcription, because these primers are degraded later in the replication sequence.

Transcription yields three types of RNA: messenger RNA (mRNA), ribosomal RNA (rRNA), and transfer RNA (tRNA).

The binding of DNA-directed RNA polymerase to the initiation site on DNA seems to help open the DNA double helix in that region. Transcription then begins as the polymerase adds NTPs sequentially in a 5′-to-3′ direction to create an RNA strand that is complementary to one of the DNA strands. RNA polymerase transcribes only one of the DNA strands at a time. RNAs may arise from either DNA strand, but the basis for the selection of which DNA strand will be transcribed has yet to be understood.

As successive nucleotides are added to the RNA chain, the 5′ end of this RNA retains its original triphosphate group, which provides one terminus of the RNA. Ultimately, RNA polymerase reaches a chain-terminating signal (not to be confused with the chain terminating codon of mRNA). The nature of this signal is still uncertain. At this point, the last nucleotide residue is added, thus providing the other terminus of the RNA.

Although the growing RNA strand hydrogen-bonds to its precursor DNA strand, this complex is not as stable as double-stranded DNA and tends to dissociate.

Actinomycin D, which is used clinically to block cell division in choriocarcinoma and Wilms' tumor of the kidney, binds to guanine in DNA and blocks transcription. Poison mushrooms contain amanitin, an inhibitor of RNA polymerase. Rifampin, an antibiotic used to treat tuberculosis, inhibits bacterial RNA polymerase. Unlike amanitin, it does not inhibit human RNA polymerase and therefore is not poisonous.

Reverse Transcription	Oncogenic (tumor-producing) RNA viruses synthesize DNA from RNA, a reversal of the usual procedure, and insert this DNA into the chromosomes of animal cells. This reverse transcription is catalyzed by an *RNA-directed DNA polymerase*.
Modification of RNA after Transcription	Following transcription, the RNA must be further modified to make it functionally active as mRNA, tRNA, or rRNA. For example, after mRNA is transcribed, a chain of more than 200 AMP residues, termed a *poly A chain*, is attached to its 3′ end. This chain apparently assists in the transport of mRNA from the nucleus into the cytoplasm, where this chain is detached.
	Transfer RNA molecules generally contain 75 to 90 nucleotides, compared to 100 to 3000 in rRNA and mRNA. Each tRNA precursor generated in transcription must be activated first by the removal of several nucleotides from each end and then by the addition of a cytidine-cytidine-adenine trinucleotide to the acceptor end of tRNA. Many of its nucleotides are altered by methylation, reduction, or sulfation to yield uncommon nucleotides such as pseudouridylic acid, inosinic acid, and methylguanylic acid.
	Ribosomal RNA binds with proteins in the cytoplasm to form the ribosomes, where additional proteins are synthesized.
THE GENETIC CODE	Each DNA strand codes for the synthesis of many polypeptides. A segment of DNA that codes for one polypeptide chain is termed a *gene*. Each mRNA can code for one or several polypeptides.

Messenger RNA is functionally, not structurally, partitioned into sequential trinucleotide fragments called *codons*. Each codon leads to the addition of one particular amino acid during *translation*, or protein synthesis. There are 4^3, or 64, possible trinucleotide sequences of the four nucleotides in mRNA. Three codons—UAA, UAG, and UGA—do not represent amino acids, but rather they signal the point of chain termination.

Most amino acids have several codons, as shown in Figure 15-1. Hence, the genetic code is termed *degenerate*; i.e., redundancy exists among the code words for most amino acids. This degeneracy usually involves the third base of the codon, although occasionally the other two bases also vary for a given amino acid. Isoleucine, for example, has three codons that differ only in the third base: AUA, AUC, and AUU. Serine, on the other hand, has six codons that exhibit degeneracy in all three bases: AGC, AGU, UCA, UCC, UCG, and UCU.

AUG, the only codon for methionine, is also the chain-initiation codon for proteins of higher organisms.

The DNA strand used as a template for a given mRNA has base triplets that are complementary to the codons (they are not considered codons themselves). Any change in the DNA base sequence will be transmitted in both DNA replication and transcription. Changes in a single base, or point mutations, may involve base transition, transversion, deletion, or insertion.

In *transitional mutations*, a purine replaces another purine in DNA and a pyrimidine replaces another pyrimidine. 5-Bromouracil, which is not used medically, is a thymine analog that can replace thymine in DNA. Since it can base-pair with both adenine and guanine, a guanine, rather than the adenine

Figure 15-1

The codon dictionary. The chain-termination codons are indicated by "End." (Adapted from A. L. Lehninger, Biochemistry [2nd ed.]. New York: Worth, 1975. Fig. 34-1.)

AAA Lys	CAA Gln	GAA Glu	UAA (End)
AAG Lys	CAG Gln	GAG Glu	UAG (End)
AAC Asn	CAC His	GAC Asp	UAC Tyr
AAU Asn	CAU His	GAU Asp	UAU Tyr
ACA Thr	CCA Pro	GCA Ala	UCA Ser
ACG Thr	CCG Pro	GCG Ala	UCG Ser
ACC Thr	CCC Pro	GCC Ala	UCC Ser
ACU Thr	CCU Pro	GCU Ala	UCU Ser
AGA Arg	CGA Arg	GGA Gly	UGA (End)
AGG Arg	CGG Arg	GGG Gly	UGG Trp
AGC Ser	CGC Arg	GGC Gly	UGC Cys
AGU Ser	CGU Arg	GGU Gly	UGU Cys
AUA Ile	CUA Leu	GUA Val	UUA Leu
AUG Met	CUG Leu	GUG Val	UUA Leu
AUC Ile	CUC Leu	GUC Val	UUC Phe
AUU Ile	CUU Leu	GUU Val	UUU Phe

that thymine would specify, can be incorporated into the complementary strand during DNA replication or transcription.

2-Aminopyrine, a drug whose medical use was abandoned years ago because of its toxic effect on bone marrow, can replace adenine (6-aminopurine) or guanine in DNA, and it, too, induces transitional mutations.

In *transversional mutations*, a purine replaces a pyrimidine in DNA and a pyrimidine replaces a purine. These replacements occur in pairs and sometimes may simply represent a transposition of the two bases; e.g., AT may be replaced by TA. Spontaneous mutations are often the result of transversions.

Both transitional and transversional mutations are *point* mutations in that they change only one codon in mRNA. A codon change from AAA to AAG, for instance, will do no harm, since both codons are read as the code for lysine. A change from AAA to ACA, however, will place threonine rather than lysine into the protein. This change in primary structure may or may not alter the secondary, tertiary, and quaternary structures of the protein or change its enzymatic behavior. In sickle-cell anemia, however, valine replaces glutamate at position 6 in the β-chain of hemoglobin, which results in severe distortions in secondary, tertiary, and quaternary structure of hemoglobin and the consequent sickling of erythrocytes.

Deletion mutations occur either as a result of the loss of a base or after chemical damage to a base, such as the alkylation of guanine with alkylating drugs. In the latter case, the damaged base cannot base-pair normally and might not be read during transcription.

Acridine causes *insertion mutations*, because it can fit between two successive bases and be read during transcription so that an additional base will be inserted into the complementary mRNA strand.

The deletion and insertion mutations, which also are point mutations, are far more detrimental than transition or transversion mutations, because they shift the triplet reading-frame. The deletion of A from the DNA below, for example, changes the triplets (not codons) from TCA, GTG, T... to TCG, TGT, ...:

$$\text{TCAGTGT}\cdots \xrightarrow{\text{deletion of A}} \text{TCGTGT}\cdots$$

All DNA triplets occurring in the chain in the 3' direction after such a *frame-shift mutation* will be thrown out of register, producing distorted mRNA codons. Frame-shift mutations near the 5' end of a gene usually produce a defective polypeptide, because most of the mRNA codons have been altered. Such mutations near the 3' end of a gene, however, change relatively few mRNA codons and may produce functional polypeptides.

Any type of mutation may produce chain-terminating codons that prematurely stop polypeptide synthesis. Double or triple base changes within a triplet are rare compared to the point mutations previously discussed.

Problem 1

What is the result of the transversion mutation in DNA from TAA to TAC (see Fig. 15-1)?

A. Codon changes from AUU to AUG; methionine replaces isoleucine.
B. Codon changes from UUA to GUA; valine replaces leucine.
C. Frame-shift occurs.
D. Inhibition of DNA replication and transcription occurs.

Problem 2

Which of the below amino-acid substitutions could result from a single base mutation in DNA (rather than a double or triple base change in the same triplet)?

A. Leu → Lys
B. Phe → Lys

C. Phe → Leu
D. Ile → Leu

Problem 3

In hemoglobin C, lysine replaces glutamate at position 6 of the β-chain of hemoglobin. Which DNA base is altered in this point mutation: A, C, G, or T?

Problem 4

Which statement about DNA-directed RNA polymerase is *incorrect*?

A. It cannot utilize NDPs.
B. It is involved in DNA synthesis.
C. It cannot synthesize RNA without a DNA template.
D. It can synthesize DNA from an RNA template.

Problem 5	Write the mRNA sequence that will synthesize the oligopeptide, Phe-Met-Lys-Trp. Although Phe and Lys each have two codons, use UUU for Phe and AAA for Lys. Now write the DNA sequence needed to synthesize that mRNA.

Problem 6	Two methods exist for repairing thymine dimers. One method uses the enzymes DNA ligase, DNA polymerase, and endonuclease, but in what order?

A. DNA ligase, DNA polymerase, endonuclease
B. DNA polymerase, endonuclease, DNA ligase
C. Endonuclease, DNA polymerase, DNA ligase
D. Endonuclease, DNA ligase, DNA polymerase

Problem 7	Place the below steps of DNA replication in proper chronological order:

A. DNA-directed RNA polymerase synthesizes RNA primer.
B. Unwinding proteins open DNA double helix.
C. DNA-directed DNA polymerase synthesizes complementary DNA strand.
D. DNA ligase joins ends of DNA fragments.
E. Endonuclease removes RNA primer.

TRANSLATION

Translation, which is defined as protein synthesis according to the amino acid code in mRNA, has four stages:

1. Amino acid activation
2. Initiation of polypeptide-chain formation
3. Chain elongation
4. Termination

Amino-Acid Activation

In activation, each amino acid has a specific *aminoacyl-tRNA synthetase* that joins it to its particular tRNA. This two-step reaction consumes two high-energy phosphate bonds, because it converts ATP to AMP and PP_i. The amino acid binds to the 3′ end of tRNA, which always has the trinucleotide sequence cytidine-cytidine-adenine.

Amino acid + ATP $\rightleftharpoons$ aminoacyl-AMP + PP_i

Aminoacyl-AMP + tRNA $\rightleftharpoons$ aminoacyl-tRNA + AMP

Net: Amino acid + ATP + tRNA $\rightleftharpoons$ aminoacyl-tRNA + AMP + PP_i

Transfer RNA has a cloverleaf shape that allows intrachain hydrogen bonding and double-helix formation in each "arm" of the cloverleaf. One arm contains the amino-acid acceptor site, while another contains a trinucleotide anticodon that is complementary and antiparallel to an mRNA codon. This anticodon on tRNA binds only to its corresponding mRNA codon. The amino acid carried by tRNA does not play a role in the selective binding of the proper aminoacyl-tRNA to its corresponding codon, but this binding allows the correct positioning of the amino acid in the sequence for the polypeptide chain.

In humans, there are two systems of ribosomal protein synthesis: cytoplasmic and mitochondrial. Human cytoplasmic ribosomes have a sedimentation coefficient of 80S and possess 60S and 40S subunits. Each subunit contains 28S or 18S rRNA plus a large variety of proteins. The mitochondria contain circular DNA and synthesize proteins in a manner closely resembling the synthesis in bacteria. Human mitochondrial ribosomes, like bacterial ribosomes, are smaller, with a sedimentation coefficient of 70S. Their subunits, with sedimentation coefficients of 50S and 30S, are also smaller than those of the ribosomes in the human cytoplasm. The 50S subunit contains 23S and 5S rRNA, and the 30S subunit contains 16S rRNA; both contain various proteins.

Initiation of Polypeptide-Chain Formation	The precise mechanism for the initiation of polypeptide-chain synthesis in humans is uncertain. In *E. coli*, three proteins, termed *initiation factors*, are required. First, the 30S and 50S subunits of the *E. coli* ribosomes dissociate. Initiation factor 3 (IF-3) binds to the 30S subunit, and mRNA and IF-1 join this complex. Next, the initiating aminoacyl-tRNA, *N*-formylmethionyl-tRNA (fMet-tRNA), binds to initiation factor IF-2. (The synthesis of every protein in *E. coli* begins with methionine.) GTP, IF-2, and fMet-tRNA join the complex of the 30S subunit, mRNA, IF-1, and IF-3, which reunites with the 50S subunit to yield an active 70S ribosome bound to mRNA and fMet-tRNA. The fMet is bound via its tRNA to the initiation codon of mRNA. GTP energizes the process via hydrolysis to GDP and P_i. The three initiation factors are then liberated from the ribosome.

Although human mitochondrial ribosomes use *N*-formylmethionine, like *E. coli* and other prokaryotes, to initiate translation, man's cytoplasmic ribosomes apparently use methionine. In the cytoplasm, mRNA binds to the 40S ribosomal subunit along with the initiation factors and GTP.

A string of ribosomes bound to a single mRNA is termed a *polyribosome* or *polysome*. Polysomes are found free in the cytoplasm of prokaryotic cells, whereas the ribosomes of eukaryotic cells are either free or bound to endoplasmic reticulum.

Polypeptide-Chain Elongation	Polypeptide-chain elongation begins with aminoacyl-tRNA binding to GTP and an elongation-factor protein. GTP hydrolysis to GDP and P_i then drives the addition of this aminoacyl-tRNA to the active site on the ribosomes.

Tetracyclines, a class of antibiotics often used in humans, bind to the 30S subunits of bacterial ribosomes and prevent the attachment of aminoacyl-tRNAs. High serum levels of tetracyclines will also inhibit human protein synthesis. The absence of even one required aminoacyl-tRNA stops elongation.

Peptidyl transferase then forges a peptide bond between the amino group of the next amino acid to be added and the carboxyl group of the previously synthesized peptidyl-tRNA that is bound to the ribosome. The resultant peptide is transferred to the incoming tRNA, and the tRNA of the last amino acid to be added is liberated. The process is repeated for each peptide bond to be formed. This peptidyl transferase reaction does not consume GTP or ATP.

197

Chloramphenicol, an antibiotic used in the treatment of several life-threatening infections, binds to the 50S subunit of the 70S ribosome of bacteria and human mitochondria and inhibits the peptidyl-transferase reaction. Erythromycin is another common antibiotic that binds to the 50S ribosomal subunits of bacteria and inhibits their protein synthesis. Puromycin, an antineoplastic agent and antibiotic, resembles an aminoacyl nucleotide. After incorporation into the growing peptide chain, it prevents further elongation. Cycloheximide, a fungicide, inhibits peptide-bond formation in humans as well as in certain fungi, but not in bacteria. The popular aminoglycoside antibiotics—gentamicin, kanamycin, and streptomycin—bind to bacterial 30S ribosomal subunits to block protein synthesis.

In the final stage of each step of polypeptide-chain elongation, called *translocation*, the ribosome advances one codon forward on the mRNA, which moves the peptidyl-tRNA from the active (aminoacyl) ribosomal site to the peptidyl site. The active site is thus cleared to bond to another incoming aminoacyl-tRNA. This translocation of the peptidyl-tRNA requires a second elongation factor and consumes one mole of GTP.

Termination of Translation	Termination is triggered when the ribosome reaches one of the three chain-terminating codons on mRNA. Several proteins, termed *releasing factors*, are also required. Peptidyl transferase cleaves the tRNA from the completed polypeptide chain.
	Four high-energy phosphate bonds are consumed to synthesize each peptide linkage: two from ATP are used during amino-acid activation, one from GTP during aminoacyl-tRNA binding to the active ribosomal site, and another from GTP during translocation.
Modification of Polypeptides after Translation	Although the initiating codon in cytoplasmic translation is for methionine, not every polypeptide retains the initial methionine, because this amino acid can be cleaved after translation.
	Certain amino acids do not have codons. In proteins, they are derived from parent amino acids after the chain is formed. Hydroxyproline and hydroxylysine, for example, are synthesized from proline and lysine, respectively, in collagen, as described in Chapter 13.
	To create disulfide linkages, the SH groups of cysteine are enzymatically oxidized and positioned to allow this bond to form after the polypeptide chain has been completed.
	Not every amino acid in humans is found in proteins; ornithine and citrulline from the urea cycle never appear in proteins.
HORMONAL CONTROL OF PROTEIN SYNTHESIS	Hormones that stimulate protein anabolism (synthesis) include insulin, the androgens, thyroid hormones, and growth hormone (STH).
	Insulin stimulates tissue amino-acid uptake and protein synthesis, while inhibiting gluconeogenesis.
	Androgens, such as testosterone, promote a positive nitrogen balance and are responsible for the larger muscle mass of males compared to females. They stimulate DNA-directed RNA polymerase and increase the rate of transcription.

Thyroid hormones and growth hormone (STH) markedly stimulate protein synthesis. When either is deficient, normal growth cannot occur.

The glucocorticoids lead to protein catabolism, because they stimulate gluconeogenesis from protein.

Problem 8

Which DNA sequence codes for the tRNA anticodon that binds to the UAG terminator codon?

A. TAG   C. CTA
B. ATC   D. GAT

Problem 9

Place the following steps of translation in the proper chronological sequence:

A. Aminoacyl-tRNA binds to GTP and elongation factor.
B. Peptidyl transferase creates a peptide bond.
C. The two ribosomal subunits reunite, bound to mRNA.
D. Amino-acid activation occurs.
E. Translocation of peptidyl-tRNA occurs.

Problem 10

Which hormone stimulates protein catabolism?

A. Thyroid hormones   D. Testosterone
B. Cortisol (a glucocorticoid)   E. Growth hormone
C. Insulin

Problem 11

Which step of translation does *not* consume a high-energy phosphate bond?

A. Translocation
B. Amino acid activation
C. Peptidyl-transferase reaction
D. Aminoacyl-tRNA binding to active ribosomal site

Problem 12

Choose the antibiotics that act by binding to the 30S ribosomal subunit of bacteria:

A. Chloramphenicol   C. Erythromycin
B. Tetracyclines   D. Streptomycin

ANSWERS

1. B. (During transcription, the DNA-triplet TAA binds to mRNA as follows:

DNA   5′ T—A—A 3′
              :: :: ::
mRNA   3′ A—U—U 5′

By convention, the polynucleotide sequence is written in the 5′ to 3′ direction. Hence, this mRNA codon is UUA, not AUU. Similarly, TAC corresponds to GUA.)

2. C, D. (A single base change converts the code for Phe—UUC or UUU—to one for Leu—UUA, UUG, CUU, or CUC. Similarly, the code for Ile—AUU or AUC—can be converted to CUU or CUC for Leu. The Leu-to-Lys and Phe-to-Lys mutations would require double and triple base changes, respectively.)

3. The two mRNA codons for lysine—AAA and AAG—replace those for glutamate—GAA and GAG. Therefore, A replaces G at the 5′ end of the codon. Hence, T replaces C at the 3′ end of the DNA triplet.

4. D. (Only the RNA-directed DNA polymerases can catalyze reverse transcription.)

5. The mRNA sequence, divided for convenience into codons, is UUU-AUG-AAA-UGG. The complementary DNA sequence is CCA-TTT-CAT-AAA (these letters given in reverse order would be in the 3′ to 5′ direction, which violates the convention of writing sequences from 5′ to 3′).

6. C. (See section on DNA repair.)

7. B, A, C, E, D.

8. A. (The anticodon is CUA, which is synthesized from the DNA triplet TAG.)

9. D, C, A, B, E.

10. B.

11. C.

12. B, D.

REFERENCES

Gelehrter, T. Enzyme induction. *N. Engl. J. Med.* 294:646, 1976.

Goodman, L. S., and Gilman, A. *The Pharmacological Basis of Therapeutics* (5th ed.). New York: Macmillan, 1975. Pp. 1287–1288 (actinomycin D), 1379 (STH), 1406 (thyroid hormones), 1456–1457 (androgens), 1481 (glucocorticoids).

Lehninger, A. L. *Biochemistry: The Molecular Basis of Cell Structure and Function* (2nd ed.). New York: Worth, 1975. Pp. 320–322, 879–881, 891–904, 912–973.

Lodish, H. F. Translational control of protein synthesis. *Annu. Rev. Biochem.* 45:39, 1976.

Weber, G. Enzymology of cancer cells. *N. Eng. J. Med.* 296:486–493, 541–551, 1977.

White, A., Handler, P., and Smith, E. L. *Principles of Biochemistry* (5th ed.). New York: McGraw-Hill, 1973. Pp. 737–774.

Index

Glucagon
  effect on glycogen metabolism,
    99–100, 118–121
  effect on lipid metabolism, 154
Glucan transferase, 78–79
Glucocorticoids
  and carbohydrate metabolism, 120–
    121
  and lipid metabolism, 154
  and protein metabolism, 174, 199
  structure, 91
Glucokinase, 106, 108
Gluconeogenesis, 113–116
Glucosamine, 80
Glucose
  in gluconeogenesis, 113–116
  in glycolysis, 106–111
  structure, 74–75
Glucose-6-phosphatase
  energetics of reaction, 66
  gout, due to deficiency, 188
  reaction, 115
Glucose-6-phosphate, metabolism, 66,
  106–108, 116
Glucose-6-phosphate dehydrogenase,
  116
Glucose phosphate isomerase, 108
α-Glucosidase deficiency, 152–154
Glucuronic acid, 80
Glutamate
  metabolism, 161–163, 172–173, 179
  structure, 2
Glutamate dehydrogenase, 162–163
Glutamine
  metabolism, 172–173, 179
  in purine biosynthesis, 183–186
  structure, 2
Glutamine synthetase, 172–173
Glutathione, 116
Glyceraldehyde, 73
Glyceraldehyde-3-phosphate, 108–109
Glyceraldehyde-phosphate dehydro-
    genase, 109
Glycerol, 65
Glycerol phosphate
  energetics, of hydrolysis, 65
  shuttle, 132–134
  structure, 86
  and triglyceride synthesis, 139–140
Glycerophosphate dehydrogenase,
  132–134
Glycine
  as a buffer, 9
  conversion to serine, 174
  in purine synthesis, 183–185
  structure, 4
Glycogen
  biosynthesis, 119–120

  breakdown, 118–119
  structure, 78–79
Glycogenesis, 119–120
Glycogenolysis, 118–119
Glycogen phosphorylase, 78–79, 118–119
Glycogen storage diseases, 115, 119–120
  and gout, 188
Glycogen synthase, 118–120
Glycolysis, 106–111
GMP (guanosine monophosphate), 99.
    See also GTP (guanosine triphos-
    phate)
Gout, 188–189
Growth hormone
  and carbohydrate metabolism, 120
  and lipid metabolism, 154
  and protein metabolism, 174, 198–199
GTP (guanosine triphosphate)
  biosynthesis, 185–186
  catabolism and salvage, 186–188
  phosphorylation, of ADP, 114
  in polypeptide synthesis, 197–198
  in pyruvate kinase reaction, 114
  structure, 99
  in succinyl-CoA synthetase reaction,
    126, 128
Guanine, 47. See also GTP (guanosine
    triphosphate)
Guanosine, 99. See also GTP (guanosine
    triphosphate)
Guanosine diphosphate. See GDP
    (guanosine diphosphate)
Guanosine monophosphate. See GMP
    (guanosine monophosphate)
Guanosine triphosphate. See GTP
    (guanosine triphosphate)
Guanylic acid, 99. See also GTP
    (guanosine triphosphate)

H. See Enthalpy
Haworth projections, 74
Heavy metals, as enzyme inhibitors, 40
α-Helix, 28
Helix, triple, 28
Heme, 130–132
Hemoglobin
  alterations in amino acid sequence,
    29–30, 194
  as a buffer, 14–15
  oxygen binding, 14–16
  sickle cell, 30, 194
Hemoglobinopathies, 29
Henderson-Hasselbalch equation, 7–8
Heparin, 80
Hexokinase, 37, 106–108
Hexosaminadase, deficiency, 152–153
Hexosediphosphatase, 114–115
Hexose monophosphate shunt, 115–116

Pyridoxal phosphate
  in heme synthesis, 131
  structure, 52
  in transamination, 162
Pyridoxamine phosphate. *See* Pyridoxal
    phosphate
Pyridoxine. *See* Pyridoxal phosphate
Pyrimidines. *See also under specific*
    compounds
  biosynthesis, 179–181
  degradation, 182
  structure, 97–98
Pyrophosphatase, 66
Pyrophosphate, hydrolysis of, 66
Pyruvate
  carboxylation, to form OAA, 114
  dehydrogenation, 48–49
  formation, from PEP (phosphoenol-
    pyruvate), 110
  reduction, to lactate, 110–111
  structure, 73
Pyruvate carboxylase, 114, 129
Pyruvate dehydrogenase, 48–49, 125
Pyruvate kinase, 110

Quaternary structure, of proteins, 29

Random coil, 28
RDA (recommended daily allowance),
    defined, 47
Reaction order, 36
Reaction potential, 67–68
Reaction rate, maximal, 37
Recommended daily allowance,
    defined, 47
Reducing sugars, 76
Reductase, defined, 34
Reduction, defined, 67
Reduction potential, 67
Regulatory enzymes, 42–43
Releasing factors, 198
Respiratory control, of oxidative
    phosphorylation, 135
Respiratory distress syndrome, 87
Respiratory quotient, 142
Retinal, 56–57
Retinol, 56–57, 90
Reverse transcriptase, 193
Rhodopsin, 56
Riboflavin, 49–50
Ribonuclease, 102
Ribonucleic acids, 100–102, 193–198
Ribose, 74
Ribose-5-phosphate, 116, 178
Ribosomal RNA, 193, 196–198
Ribosomes, 193, 196–198

Rickets, 58
RNA (ribonucleic acid),
    100–102, 193–198
RNA-directed DNA polymerase, 193
RNA-directed RNA polymerase, 191
RNase, 102

Sanger's method, 26
Scurvy, 54
Secondary structure, of polypeptides, 28
Semiconservative replication, DNA, 191
Serine
  conversion to glycine, 174
  destruction by acid hydrolysis, 25
  role in cysteine synthesis, 170–171
  structure, 4
Serotonin, 169–170
Sickle cell anemia, 30, 194
Soaps, 86
Somatotrophic hormone. *See* Growth
    hormone
Specific dynamic action, 161
Sphingolipidoses, 152–154
Sphingolipids
  catabolism, 152–154
  structure, 88–89
Sphingomyelin
  accumulation, in Niemann-Pick
    disease, 152–153
  structure, 88
Sphingomyelinase, 152–153
Sphingosine, 88
Squalene, 149–151
Standard state, 63
Starch, 78–79
Steroid, structure, 90–92
Sterols, 92
STH (somatotrophic hormone). *See*
    Growth hormone
Substrate-binding, to enzymes, 35–36
Succinate, 126–128
Succinate dehydrogenase, 126, 128,
    131–134
Succinyl-CoA, 126–128, 143
Succinyl-CoA synthetase, 126, 128
Sugars, 73–80
Sugars, reducing, 76
Sulfanilamide, inhibiting folate
    synthesis, 41
Sulfatase, 34, 152–153
Sulfatidase deficiency, 152–153
Sulfatides
  in metachromatic leukodystrophy,
    152–153
  structure, 89
Surfactants, 87
Synthase, defined, 34